FENDER AMPS
THE FIRST FIFTY YEARS

by **JOHN TEAGLE**
& **JOHN SPRUNG**

Edited by Jon Eiche

A Joint Publication of

HAL•LEONARD CORPORATION
7777 W. BLUEMOUND RD. P.O. BOX 13819 MILWAUKEE, WI 53213

Copyright © 1995 HAL LEONARD CORPORATION

All rights reserved. No part of this book may be reproduced or transmitted in any form or by any means, electronic or mechanical, including photocopying, recording, or by any information storage and retrieval system, without permission in writing from the publisher.

Published by HAL LEONARD CORPORATION
7777 West Bluemound Road
P.O. Box 13819
Milwaukee, Wisconsin 53213

Printed in Milwaukee, Wisconsin

ISBN 0-7935-3733-9 (soft cover)
ISBN 0-7935-4408-4 (hard cover limited edition)

Library of Congress Cataloging-in-Publication Data:

Teagle, John.
    Fender amps: the first fifty years / by John Teagle & John Sprung; edited by Jon Eiche.
       p.   cm.
    Includes bibliographical references and index.
    ISBN 0-7935-4408-4 (hard cover w/slip case). — ISBN 0-7935-3733-9 (soft cover: alk. paper)
    1. Fender Musical Instruments—History. 2. Guitar—Electronic equipment. 3. Audio amplifiers.   I. Sprung, John. II. Title.
ML1015.G9T42 1995
681'.867—dc20   95-4466
                                                                               CIP
                                                                               MN

It's what's inside that counts. *Readers waiting for* Fender Amp Boxes: The First Fifty Years *will have to be content with this photo of vintage shipping cartons.*

TABLE OF CONTENTS

Foreword 7
Preface 8
Acknowledgments 10
Photo Credits 12

Part I: History
Fender Company History 16
Fender Amp Lineage 24

Part II: The Amps

Tube Amps
Deluxe 1945–1985, 1994–Present 42
Princeton 1945–1985 48
Pro 1946–1982 53
Super 1947–1982, 1988–Present 59
Champ 1948–1994 66
Bassman 1952–1983, 1990–Present 70
Twin 1952–1985, 1987–Present 80
Bandmaster 1953–1980 89
Tremolux 1955–1966 94
Harvard 1955–1961 97
Vibrolux 1956–1982, 1995–Present 99
Vibrasonic 1959–1963, 1972–1981, 1995–Present 102
Concert 1959–1965, 1982–1985, 1993–Present 106
Showman 1960–1981, 1987–1993 110
Vibroverb 1963–1964, 1990–Present 115
Bronco 1967–1975 118
Bantam Bass 1969–1971 120
Musicmaster Bass 1970–1982 121
400 PS Bass 1970–1975 122
Super Six Reverb 1972–1979 125
Quad Reverb 1972–1979 126
Super Twin 1975–1980 127
300 PS 1975–1980 130
Studio Bass 1977–1980 131
75 1980–1982, 30 1980–1981, 140 1980 132
RGP-1 and RPW-1 1982–1984 135
Tweed Series 1993–Present 137
Custom Amp Shop 1993–Present 139
New Vintage Series 1995–Present 142

TABLE OF CONTENTS

Solid-State Amps
First-Series Solid-State 1966–1971 144
Super Showman 1969–1971 146
Zodiac Amps 1969–1971 148
B-300 1980–1982 149
Second-Series Solid-State 1981–1987 150
Third-Series Solid-State 1986–Present 152

Effects
Volume Pedal 1954–1984 156
EccoFonic 1958–1959 158
Reverb Unit 1961–1966, 1976–1978, 1994–Present 159
TR 105 1961–1962 162
Electronic Echo Chamber 1963–1968 163
Solid-State Reverberation Unit 1966–1972 164
Echo-Reverb 1966–1970 165
Soundette 1967–1968 166
Vibratone 1967–1972 167
Orchestration + 1968–1970 169
Dimension IV 1968–1970 170
Fuzz-Wah 1968–1984 171
Multi-Echo 1969–1970 172
Fender Blender 1968–1977 173
Phaser 1975–1977 174
Miscellaneous Stomp Boxes 1986–1987 175

Fender Amps Family Portrait 176

Part III: Color Gallery 177

Part IV: Ampology
Amp Basics 218
Maintenance and Mods 227

Part V: Appendices
Parts 232
Availability Charts, with Prices 238
Dating Your Fender Amp 244
Chronology of the Earliest Fender Amps 248
Bibliography 251

Index 253

About the Authors 256

FOREWORD

For the last 50 years, most unforgettable musical performances have usually come from talented musicians with amplified instruments. Specific to rock, jazz, blues, or country, often those musicians have chosen a Fender amp to support their sound.

The book you hold tells one of the most significant stories in the music industry: the creation, evolution, and continuation of a very relied upon and relished instrument, the Fender amplifier.

From its birth in 1945 through today, the legacy continues to have a successful impact on modern music. This reputation is one to admire and honor. Also, it is a great source to build upon.

Today, Fender Musical Instruments continues to dedicate its people and resources to instrument amplification. As we expand our efforts we strive to sustain the quality and innovation that are so well researched and presented in this book.

The following presents a special era in amp design and performance. On behalf of Fender employees past and present, I welcome you to get better acquainted with this part of music making history.

WILLIAM C. SCHULTZ
President,
Fender Musical Instruments Corp.

PREFACE

What you're holding is a record of all the different versions of all the different amps bearing the Fender name. For a glimpse of the range of music-generating boxes spanned in these pages, take a look at the Color Gallery (if you haven't already peeked), on pages 177–216.

Because of the amount of ground this book needs to cover, it's divided into several parts, each of which can be read on its own. Part I gives a historical overview, first of the Fender company in all its incarnations, and then of the amps and the changes that have taken place over the years that are common to the whole line: coverings, grilles, knobs, etc. This "Lineage" chapter also acts as a guide to Part III, the Color Gallery, which is likewise arranged chronologically, starting with the 1945 K&F amps and extending to today's top-of-the-line models from the Custom Amp Shop.

Between Parts I and III is the main section of the book, laid out in what hopefully is the most useful and logical arrangement possible. All the amps are presented in their entirety, one by one in separate chapters, from their introduction to their discontinuation. It was decided to treat each amp as if it were the most important model and do a thorough year-by-year look at each variation. If you have one of these amps and want to know more about it—as well as other versions from different eras—head straight to that chapter. Some redundancy between chapters was necessary, so if you read this book from beginning to end you may encounter a bit of repetition. Where feasible, a cross-reference is made to the amp on which a feature is first seen (e.g., piggyback amps—see Showman) for a full discussion of that feature. Because changes occurred on different models at different times, or in some cases not at all, charts would have been incapable of showing the whole picture.

Part IV addresses electronics. "Amp Basics" covers how the circuitry of Fender amps advanced over the years, while "Maintenance and Mods" presents a simple, commonsense approach to those two topics. In a perfect world, each of these chapters could have benefitted from book-length treatment on its own.

Closing out the book is a collection of appendices that includes a pictorial overview of the parts found on Fender amps; a series of charts showing what was available when, and for how much; a guide to determining how old your amp is; and a bibliography.

As you read through the chapters in Part II, you may run across some *italicized* terms (*wide-panel* cabinet, *long-tailed* phase inverter, etc.). These are clarified elsewhere in the book. "Fender Amp Lineage," in Part I, covers terms related to the cosmetics of the amps, while "Amp Basics," in Part IV, deals with electronic terms.

A word about how the co-authors divided their duties: In simplest terms, John Teagle did the writing (references in the text to "the author" refer to him), and John Sprung took the photographs; both participated in the necessary research. Let "the author" exercise the power of the pen in giving proper credit to his cohort: First of all, Mr. Sprung did an outstanding job of putting together just about every amp the company made in its first 20 years, plus a two-page look at all the different styles from the last 50 years (the foldout "Fender Amps Family Portrait" at the beginning of the Color Gallery). Hunting down all the pieces was a major undertaking; putting them all together in one room deserves a round of applause from the public and a tip of the hat from me. Second, he has closely examined more early Fender amps than anyone, and his knowledge of the subject is responsible for the thoroughness of the discussions regarding them that are found in these pages.

One more tip of the hat: to the present management of Fender Musical Instruments Corp. Had they not been so dedicated to turning the company around—particularly the amplifiers—writing this book could have been a very depressing project. It would have been sad to have written about the history of the amp line 10 years ago, when production ground to a halt. Instead, the continuous rise of quality in design, execution, and variety of models since that time points to a happy future for the King of Amplifiers.

ACKNOWLEDGMENTS

From John Teagle:

John Sprung (and family), thanks for getting me involved in the project and for going beyond the call of duty in helping every step of the way.

Thanks to all the others involved in bringing the idea to print: Michael Purkhiser, of Purkhiser Electronics, Akron, Ohio, for continually advising me on the technical aspects. Blackie Pagano of Tiger Tech, and Louie Rini, for additional advice. Dan Courtenay of Chelsea Second Hand Guitars, NYC, for helping out with time and money, as well as the unlimited use of his store as a photography studio, mail room, meeting place, and examining room; also Les Leiva, Bobby, Gonzo, Eric, Danny, Dennis, Doug, and all the friends of CSHG. John Peden, Brian Fischer, Gil Southworth, Buck Sulcer, G.E. Smith, Bill Victor, Don Zesiger, Plummer Munt, Tom Verlaine, Louie Hrusovsky, Chuck Lauricella, Gordon and Paula Dow, Mark Benson, Artie Smith, Larry Wexer, Roger Thurman, for the use of all the lovely amps. Thom and Kathy Humphrey, of Ross Music, for being such kind folks. Roger Weston, for his photographic skills and patience. Fred Popovich, Kevin Macy, Bob Ohman, William Baker, Jim Colclasure, Jim Bollman at Music Emporium, Doug at City Guitar, Steve Soest, Ryland at Rockaholics, Jimmy and Stevie at Outlaw Guitars, Chris at Mojo (NYC), Nate at Mojo (Napa), Peter Wentz, Dave Hussong, Jimmy Brown, Stan at Elderly, Bob at Park Avenue Records, Richard Smith, for catalogs and information. Ken Feist, longtime Fender rep and pal from my days at Henry's Music; also Paul Teagle, Danny Basone, Henry Pawlak, Bernie Michael, Don Dixon. Laura Evans, Libby Braden, Shelly Gargus, for typing and deciphering. Brad Smith and Jon Eiche, from Hal Leonard, for their patience and dedication to the project; also their wonderful families. Mike Lewis, Dwight Doerr, Steve Grom, from Fender. Jay Scott and Ritchie Fliegler, for authorly advice. Chris, Ted, Baker, Pete, Claudine, Doug, Mike, Bob, and Hammer, for trying to play in a band with someone whose mind is elsewhere. Thanks especially to all the family, in particular Phil and Mary Teagle, for allowing me the time to research things, and the Evans', for their hospitality.

Finally, I'd like to dedicate this book to Melanie and Cassidy Teagle, for donating to this project the use of their husband and father, respectively, for the last year.

ACKNOWLEDGMENTS

From John Sprung:

Thanks to all of the following good people (in alphabetical order) for help in preparing this book: Jeff Bober; Alex Breger; Barry Brody, Rehearsal Space; Craig Brody; Dave Brown; Mike Colburn; Dan Courtenay; Jim Crenca; Cesar Diaz; André Duchossoir; Ed Eastridge; Jan Faul (for assistance with 8x10 photography); Brian Fischer; Stephanie Fischer; Ritchie Fliegler; Justin Galenski; the late Danny Gatton; Alan Hardtke; Steve Hash, Photopro; Paul Herman; Sam Hutton; Nate Kramer, MoJo Music; Buzz Levine; Mike Lewis; Victor Lindenheim; Nils Lofgren; Van Lurton; Steve Melkisethian; Mark Mitchell; Albert Molinaro; Jay Montrose; Plummer Munt; John Peden; Aspen Pittman; Red Rhodes; Joe Romero; Bill Ryan; Paul Schien; G.E. Smith; Richard Smith; Gil Southworth; Buck Sulcer; Technical Photo; Mike Toperzer; Washington Music and the Levine family; James Werner; Phill Zavarella; Bruce Zinky.

Special thanks to Jenny, my wife, for putting up with "the last photo shot" over and over again.

PHOTO CREDITS

The following people have contributed amps and guitars for the many photos appearing in this book. Without their donations, this book would not be possible.

Brian Fischer: Deluxe #A01845; Echo-Reverb #10892; Musicmaster Bass #A-758603; Dimension IV #005519; Champ #C-22125; Champ #A-07669; Vibro Champ #A18428; Black Reverb Unit; Concert #A00833; Princeton Reverb #A04395; Pro Reverb #A08197; Vibrolux Reverb #A0455; Princeton #P00145; Bassman #BM02028; 1948 Princeton; Deluxe "Model 26" #749; Champion "800" #591; Super, slant-front, #910; Super Reverb #A26196; Bronco #A26196; Tremolux #A02433; Bassman, piggyback, #A31227; Deluxe Reverb; Bassman head #BP09677; Vibroverb, brown; Super, tweed; Twin, tweed; and numerous other amps that were used for close-up shots. Brian was also kind enough to supply us with original banners, signs, and literature. Brian and Stephanie opened their home to repeated invasions for yet more photos. Special thanks to Popeye.

Alan Hardtke: Wood Professional #1137; K&F, 8"; Showman head and cab #01486; Bandmaster head and cab #54512; Bassman cab #M03399; Tremolux cab #L00346; White Amp and Steel #AS00243; Tremolux head #00469; Yale amps; Super, slant; Bassman, 1952, prototype; Princeton, wood; "Model 26" #847; Deluxe Reverb #A02047; Princeton, wide-panel; Harvard #H00239; Bandmaster, 3x10, #00447; Dual Showman cab, white; Vibratone #3144; tweed extension cab; 2x12 cab; many close-up shots of amps too abundant to mention. Thanks to Alan for the use of his catalog collection and for putting up with all the hardships of owning so much cool stuff.

Gil Southworth: Twin, white, #00178; Twin, white, #00159; Pro #54013; Deluxe #D00760; Concert #56849; Vibrasonic #00481; Reverb Unit #R04375; Reverb Unit #R02570; Reverb Unit #R00177; "Model 26" #903; Concert #52139; Reverb Unit #R04100; Vibroverb #00444; black extension cab, 12" and 15"; Twin Reverb; Super #02917; Concert #01737; Vibrolux #01636; Pro #S01477; Tremolux #01839; Twin #A0345; Bandmaster #S01210; Super #00736; Princeton #P04234; Deluxe #6769; Super #5002; Princeton #2895; Twin #0446; Pro #1104; Esquire guitar, 1957; Strat, 1965; Strat, 1957; cardboard box pile.

PHOTO CREDITS

John Sprung: Dual Professional; Deluxe, 1948; Pro, 1948; Champion "800"; "Model 26"; wood Princeton; wood Professional; K&F steels; Fender steels.

Buck Sulcer: Champ #13417; Princeton #1775; Princeton #4273; Vibrolux #F4004; Deluxe #1879; Deluxe #D01676; Vibrolux #01788.

Dan Courtenay: 1x15 Bassman; 1x15 Bassman; Dimension IV; Dimension IV Universal and Fender curtains; Super ChampDeluxe.

Bill Victor: Bandmaster, 1963.

Don Zesiger: Bassman, 1964.

Craig Brody: Vibrasonic, 1963, white Tolex; Concert, 1963, with tags.

Ed Eastridge: Vibroverb, 1964.

Jim Crenca: Champ, 1953; Bassman 20.

Mike Colburn: Fender sign.

Joe Romero: Super Champ.

Mike Toperzer: Twin; Quad Reverb.

Aspen Pittman: "Model 26"; K&F.

Danny Gatton: P-Bass, 1958; Custom Shop Doubleneck; Custom Shop Tele.

Nils Lofgren: Princeton Reverb.

Albert Molinaro: K&F, 10".

Chuck Levine and family: all of the new amps.

Justin Galenski: Princeton Reverb.

Buzz Levine: Champ.

G.E. Smith: P-Bass, 1951.

Phill Zavarella: Zodiac.

Plummer Munt: Bantam Bass.

Van Lurton: brown Reverb Unit.

Victor Lindenheim: Tele, 1953.

Special thanks to: Roger Weston, for color photos, pages 196 and 197; Richard Smith, for providing previously unpublished factory photos taken by Leo Fender; The Bold Strummer Ltd., for permission to reprint the photo of the Bob Wills custom K&F amp.

PART I: HISTORY

FENDER COMPANY HISTORY

Beginnings

What is known today as Fender Musical Instruments Corp. has its roots in a time a half century ago when technological, social, and musical changes converged to make electric guitars and amplifiers practical, affordable, and desirable.

Clarence Leo Fender, born in southern California in 1909, was a self-taught electronics buff who grew up with the first vacuum tubes and radio broadcasts, finally establishing his own repair shop—Fender's Radio Service—in 1938. The shop was home to a number of endeavors: renting homemade P.A. systems for social events, selling records and musical instruments, and, of course, repairing radios and other electrical devices. During WWII Leo teamed up with a musician and instrument designer named Clayton Orr Kauffman, known to one and all as Doc, to design and eventually manufacture amplifiers and electric lap steel guitars featuring a pickup Fender had developed. Their enterprise, launched in 1945, was called K&F (Kauffman and Fender).

This was a time when both Hawaiian and Western Swing music flourished, and the market for K&F products looked promising. Soon Leo wished to expand the business and build a separate manufacturing facility. Doc had cold feet and bowed out by February of '46, leaving Leo with four employees, a somewhat established business, and high hopes. Fender Electric Instrument Co. was born.

Expanding as planned, Leo had a pair of unheated, un-air-conditioned sheet-metal buildings erected later in the year, providing the business with 3,600 square feet of space in which to operate. This was an enormous amount compared to accommodations at the radio shop, where Leo still worked during the day. This arrangement would not last much longer.

PART I: HISTORY—*FENDER COMPANY HISTORY*

The Radio-Tel Years

Radio & Television Equipment Co. was an electrical supplier operated by F.C. Hall (later of Rickenbacker). Their salesman handling Fender's Radio Service was Don Randall (later of Fender Sales), who convinced the two parties to enter into a distribution agreement for Fender's guitars and amplifiers during the second half of '46. While Leo would later complain about a lack of promotion during their early days, Radio-Tel did print up a small foldout brochure that featured "AMPLIFIERS In beautiful woods to match guitar models. Three Sizes - Chrome Trim." Actually, "featured" is an overstatement, as almost 95% of the brochure's space was devoted to the three single-neck guitars.

Times were not good at first for Leo's business, which survived off the radio shop, as well as his wife's income. In former Fender employee Forrest White's book *Fender: The Inside Story*, Leo is quoted as saying, "Those years were absolute hell. I think I worked from six in the morning 'til midnight every day of the week. A new trademark is a hard thing to get accepted. With no advertising, no one knew who we were and there was nothing to pep up sales. It took every penny I could get my hands on to keep things together."

By the late forties, as more and more artists began using Fender equipment, word of the company spread. Business picked up and sales personnel placed merchandise to all parts of the country. Around the beginning of '48, Dale Hyatt, who had worked at the radio shop and on the assembly of many of the early Fender products, bought the repair business from Leo, who now devoted all his time to guitars and amps. A third building was added c. late '49, increasing the company's floor space by about 50%. While not opulent, it was constructed of brick, with tiled office space and bathrooms. Approximately 15 people were on the payroll by this time, including George Fullerton, who would be closely associated with Leo for the next 40 years! Much of the early wiring was done by young ladies Lupe Lopez, Maybelle Ortega, and (later) Lydia Sanchez. Louis Lugar built the cabinets, and Ray Massie, a long-time associate of Leo's at the radio shop, helped design some of the early circuits.

Amp assembly, 1950. *Lupe Lopez, installing caps, resistors, and AC cords in Champs.* (Photo by Leo Fender, courtesy of Richard Smith)

The tweed-covered, chrome-chassis amps gained the reputation of being nearly indestructible on the road, as well as being the most powerful on the market. The Champion line was on its way to a long-term relationship with students and teachers, and the multi-neck steel guitars (including the three-neck Custom, released in '49) had become the choice of professionals across the Southwest. The well-informed in the rest of the country were also becoming aware of the relative newcomer. As the forties came to a close, Fender had become a major player in the field, advertising regularly in the national trade magazines and establishing dealerships across the country. The trend of phenomenal growth had begun.

The amplifier line remained stable during the first part of the fifties as efforts were put into new instruments. The Esquire/Broadcaster/Nocaster/Telecaster (which had its origins in a patent obtained earlier by Leo and Doc) settled in by the spring of '51. Later that year another eye-opening instrument was released: the Precision Bass—arguably the most important contribution to music that Fender has made. A new amp, specially designed to reproduce the bass, was released the following spring; ironically, the Bassman would become one of the greatest *guitar* amps of all time. Later that year another new amp, the prototype high-powered 2x12 Twin, was unveiled. With work on what would become the Stratocaster underway, Fender obviously had the recipe for success. The following year would bring great changes to how the company operated.

Fender Sales

Up to this point, three of the most important people involved with the Fender line were neither employees nor business partners, *per se*. Don Randall's presence was probably more crucial to the company than that of anyone on the payroll, yet he was working for F.C. Hall's Radio-Tel in Santa Ana, as was ace salesman Charlie Hayes. A "new order" entailed the ending of Radio-Tel's exclusive distribution deal, with a replacement company set up to perform the old duties but share evenly in the profits. With a capital investment of $25,000 each, Randall, Hayes, Hall, and Fender formed Fender Sales in 1953 to promote and distribute the line of products manufactured in Leo's solely owned factories. Randall would work closely with the R&D

department, offering suggestions as to what products they should be designing and whether the ideas R&D was coming up with had any marketability. His role would become even greater a few years later.

New offices were set up for Fender Sales, and four new factory buildings encompassing 20,000 square feet replaced the old sheet-metal pair. The land had been purchased the previous year and included several acres, allowing for possible expansion if things continued to grow. The next dozen years saw not only the growth in market share that was optimistically predicted as the new-fangled instruments became more accepted, but also the growth of the industry as a whole in a manner no one could have envisioned. With new R&D man Freddie Tavares as Leo's right arm, and new foreman Forrest White forcing a well-organized factory on the old-timers, Fender made yet another giant addition to its line as the Stratocaster was completed in April of '54.

Larger quarters, 1953. *Champs, Princetons, Deluxes, and Pros await final inspection before donning their back panels. (Photo by Leo Fender, courtesy of Richard Smith)*

Charlie Hayes died in a car crash the following year, and things changed again in the upper management. Randall and Fender decided to buy out both Hayes' widow *and* Hall, who had bought Rickenbacker not long after becoming a partner in Fender Sales. The Rick line was being sold through Radio-Tel, just as Fender had been from '46 through '53, only in this case Hall owned the factory. Somewhat understandably, he had not been privy to all the goings-on at Fender, and by the end of '55 he was no longer a part of the company.

Business was still growing, and the work force was up to 50 employees. The following year two more buildings were added, which before long would not be sufficient, and in '59 three more were added, bringing the floor space up to 54,000 square feet—over 10 times the amount the company had started the decade with. Production figures and total sales went up even more. Employment reached the triple-digit mark! The business, as well as the industry as a whole, continued its seemingly endless growth, with the company expanding into acoustic guitars in '63. Space away from the factories was leased for warehouses and the new acoustic-guitar production line, and a short time later more was leased for building amplifier cabinets.

Leo and Freddie, c. 1956. *A familiar sight—Leo Fender (seated) and Freddie Tavares working together. Here they're testing a Tremolux. (Courtesy Hardtke Archives)*

By the end of '64, close to 600 people were employed, working out of 17 (by some accounts as many as 27) buildings covering over 100,000 square feet. The line of products was still growing, as were sales and production figures. Apparently the market had no limit; but Leo, who was in poor health, was overwhelmed and wanted out. Don Randall saw to it that he got his wish, as well as a considerably larger cash settlement than expected. CBS, foreseeing continued growth, paid a record $13 million for everything.

CBS

CBS had their own ideas about a number of things, first of which was the factories. It was decided there would be no more separate leased buildings, and a huge new plant was built to house everything. One hundred seventy-five thousand square feet were to be for manufacturing, with 55,000 square feet more for offices, etc.—five acres under one roof! CBS execs were confident of this approach, assuming things would maintain or grow. Leo had always kept the size of his buildings down for fear of one day having to sell them, possibly one at a time.

FENDER FACTORY NEARS COMPLETION

The new 175,000 square foot manufacturing facility now under construction in Fullerton, California as seen from the air gives an idea of the expansion going on at Fender. Adjacent to the new plant are nine buildings of the old facilities which will be retained upon completion. Ten other buildings at six loctions will then be amalgamated into the new plant. Present schedule for occupancy—April 1st.

CBS expansion, 1966.

Leo would continue to work for the R&D department following a short leave of absence, but his ideas no longer carried much weight. A new office/lab for his last years at the company ('65–'70) was set up out of the way, as the new CBS folk worked to fit the company into the larger corporate picture.

"Transistorized" must have seemed like the direction to head in as the new company looked to progress. After all, the old company had succeeded by continually upgrading existing models and introducing new ones. Fender had been a progressive company for years, and CBS was staying to the plan. But their efforts, however noble, were not given the proper time to develop. Coupled with that was the "fact," unbeknownst to those involved, that solid-state amps simply don't hold up against tube amps; they were pursuing something that wasn't there.

Nearly 30 years later it's easy to see the error of their ways, but in the mid sixties transistors were seen as the keys to the future. Paul Spranger, Dick Evans, Bob Rissi, and Seth Lover (of Gibson P.A.F. and Fuzztone fame and inventor of Fender's first humbuckers) were some of the people involved in the first wave of solid-state amps, which later included the Zodiac models. In '71 they gave up. Despite the initial solid-state fiasco, CBS would continue to build a variety of tube amps. Efforts for the next 10 years would go towards improving the standard models and releasing new designs at the top of the line. In the long run, Fender's amp line survived due to the timelessness of the pre-CBS designs—the all-tube blackface combos and piggybacks. Minor changes to their cosmetics (silverface, blue sparkle grille, etc.) and circuitry occurred, but CBS' progressive ideas would wisely be reserved for new models. The 400 and 300 PS amps, the Super Twin, and the 75, 30, and 140 all featured totally new designs, both electronically and cosmetically. Chief engineers during this period included Ed Jahns and Bob Haigler. Freddie Tavares was still the company's musical right arm.

Freddie Tavares *(left)* **and Ed Jahns, 1975.**

Sales on the standard silverface amps finally started to dwindle, although a full line was still offered nearly 20 years after their initial release. A new trio of high-gain amps with channel switching—the 75, 30, and 140—was released in 1980 in response to the number of amps from competitors offering a sound that wasn't available from the silverface models. These were replaced by a line of similar amps conceived by newcomer Paul Rivera featuring the look of the mid-sixties line. The silverface amps were discontinued, and a new line of solid-state amps was offered, this time consisting of a notably better product. These amps, along with the new line of reissue guitars, helped Fender regain credibility during the last years of CBS ownership. Nevertheless, Fender struggled with a changing market and many internal changes. One such change: moving amplifier production to the Gulbransen Organ plant in Hoopeston, Illinois, where large amounts of production time were starting to become available due to declining sales of home organs.

FENDER AMPS: *THE FIRST FIFTY YEARS*

FMIC

Although Fender continued to make money for CBS, most of the companies in the Musical Instruments Division did not. CBS decided in 1984 to sell them all, including Fender. In early '85 a group of Fender's top brass and "foreign distributors" fronted a $12.5 million bid for basically the rights to use the Fender name. The sales network remained, but the new company had to start its manufacturing from the ground up, in brand-new locations. Imported guitars and solid-state amps, as well as leftover CBS inventory, had to suffice until new guitar and amp factories could begin operations.

In late '85 the new Fender Musical Instruments Corp. (FMIC) bought the Sunn company, which at one point had been an important maker of guitar and bass amps, but had more recently concentrated on P.A. and lighting systems. Within six months operations were moved from the old Sunn factory in Tualatin, Oregon, to a 25,000-square-foot facility in nearby Lake Oswego. There Fender not only resumed production of Sunn equipment, but also began making new guitar amps under its own banner. Roger Cox, Vice President of Research and Development, set about enlarging the R&D department to include the design of several tube and solid-state guitar amps, circuit design and packaging for the imported Sidekick amps, development of new pro audio products, and improvement or replacement of several Sunn products. The department was enlarged in all areas—engineers, technicians, draftsmen/designers, and several outside consultants. Some of the people involved in those early days of FMIC were Marketing Director Steve Grom, Bob Haigler, Bill Hughes, Mark Wentling, and Cal Perkins. Among the first products for the new company were "The Twin," the Dual Showman, the Stage 85, the 2235, the Sidekick 100 Bass, and re-do's of several Sunn models.

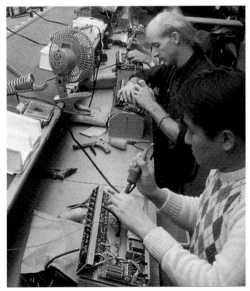

Lake Oswego, 1989. *Workers assembling Champ 12 chassis.*

Both tube and solid-state amp lines have expanded repeatedly, and so have the facilities. The original amplifier plant initially encompassed both factory and warehouse; in July 1993 the warehouse moved across the street so the factory could expand. The Custom Amp Shop was launched that year. Today Fender employs 190 people in Lake Oswego. A new factory for solid-state amps recently opened in Corona, California, adjacent to the Fender guitar plant.

Research and development has continued to grow and today is based at company headquarters in Scottsdale, Arizona.

As the Fender amp approaches its fiftieth anniversary, over 30 models are now offered in eight different series. Tube amps make up more than half of the line for the first time in years—an encouraging figure to fans and devotees of Fender tube amps. Sales are on the rise, and the Fender reputation has been polished up, ready to shine brightly for another 50 years.

FENDER AMP LINEAGE

K&F, c. '45. This model has an 8" speaker.

Back of 1x10 K&F. Only a handful of K&F amps have survived. But these ubiquitous few have appeared in numerous books and magazines, giving the impression of multiplicity.

Inside the chassis of a 1x10 K&F. Note lack of circuit board.

Bottom of a K&F. The bare wood is where the cabinet was mounted while being sprayed with the "crinkle" finish.

Fender Amplifiers...these two words would be synonymous, had Leo Fender not also made guitars and basses. To some people, including the author, they go hand in hand, the thought of plugging an electric guitar into anything but a Fender being a frightful one. But it's not just *tone* that make these amps so appealing (although this should always be a priority). Fender amps are cultural icons, having been used on a large portion of the great records for the last half century, from Wes Montgomery's rich octaves to Paul Burlison's overdriven rockabilly and everything in between. Domination in styles as varied as electric blues and surf instrumentals shows their inherent versatility. Another reason is that they always looked great, whether a matching set of Beach Boys white Tolex amps or a solitary blackface Super Reverb. A tweed Deluxe is becoming acceptable living room "furniture" in more and more households these days—a functional work of art. The fact that these styles, some of them 40 years old, are still being used by Fender (and generally considered to be the best looking in the business) is a tribute to their original design.

This chapter covers the general changes to the amp line as a whole over the last 50 years, including cosmetic changes such as covering, grille cloth, knobs, handles, logos, control panels, etc. The Color Gallery (pages 177–216) parallels this chronological overview, offering at least one example of each different "style" or combination of features; cross-references to that section are found throughout this chapter. Changes specific to the individual amps, and details on how they work, will be covered in detail in Part II of this book, which devotes a separate chapter to each model, following it from beginning to end.

Although Leo had built a number of homemade amps and public address systems in the early forties, the earliest production amps were the K&F models of 1945. The individual amps had no specific names and were basically handmade. A model with a single 8" speaker, one input, and a Volume control, and one with a 10" speaker, two inputs, and controls for both Volume and Tone were primitive in their construction and finish. At least three different styles of knobs were used. A gray crinkle finish was sprayed onto the cabinets, with bare wood showing on the bottom where a bracket held the box during spraying. (It's said that this finish was baked on in Mrs. Fender's oven.) A metal mesh grille was unique to the K&Fs. The chassis was bent steel, with no markings for the controls. A metal badge was affixed to the front, above the speaker grille, with the company's logo. Field-coil speakers, which required an electric current to create the necessary magnetic field, were standard fare at the time. A heavy-duty leather handle completed the package. (See page 177 for color photo.)

A late K&F model. *This was reportedly custom-made for Bob Wills and his Texas Playboys.* (Photo courtesy of The Bold Strummer Ltd.)

After Doc Kauffman left the project in 1946, Leo quickly came out with a new line of amps, all in hardwood cabinets. "Gleaming blonde maple, black walnut, and dark mahogany" were the official choices, with examples of oak showing up on occasion. A matching wooden handle was secured to the top. Three different sizes of these "woodies" were offered: the 1x8 Princeton, the 1x10 Deluxe (which was labeled "Model 26"), and the 1x15 Professional. The Professional and Deluxe were equipped with both mic and instrument inputs, and pointer knobs were standard on all the amps. The grille cloth varied in color; red, blue, and yellow/gold were standard. A panel at the bottom rear of the cabinet was covered in matching material. Three protective metal strips were mounted over the speaker opening, "which add flash and brilliance to their already sparkling appearance." A new logo, with the old lightning bolt and the words "Fender Electric Instrument Co./Fullerton California," was stenciled on the new control panel, along with markings (1–12) for the controls. The Princeton did not have this panel and, like the K&Fs, had no markings on the chassis. (See page 178.)

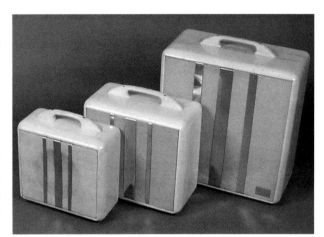

Fender "Woodies," c. 1946. *Complete line of amps. L to R: Princeton, Deluxe, Professional.*

A new style of cabinet was released c. 1947 that would be the basis of all the amps though 1959. The Dual Professional amp introduced "highest grade airplane luggage type linen fabric" covering, later referred to as tweed. The first version was a uniformly colored weave, applied straight up and down. There was no finish to the material. Complementing the light color of the covering were a dark brown, fuzzy grille cloth and a dark brown leather handle. The chassis was mounted on the top piece of the cabinet, with a new chrome-plated control panel facing up instead of back. Unique in the line of Fender amps was the metal brace that ran down the middle of the amp, holding the two baffle boards together. An embossed metal plate with "Fender" in block letters and the words "Dual Professional/Fullerton, California" was mounted on the front, above the brace. The amp introduced the On/Off switch, the panel-mounted fuse holder, and the "bull's eye pilot light," as well as finger-joint cabinet construction and the use of circuit boards. (See page 179.)

The change from wooden cabinets to tweed-covered ones seems to have happened this way: The Dual Professional appeared by early '47, introducing tweed to the Fender line. This was the monochromatic vertical weave. The wooden Professional then became the tweed Pro, c. early '47. By this time the vertical tweed contained threads of two contrasting shades. The wooden Deluxe continued to sell into April of '48, probably to use up existing parts, before changing to tweed. (For a detailed look at this subject, see page 248.)

Except for the V-front Dual Professional (which became the Super), the tweed amps initially had *TV-front* cabinets, so called because of the rounded corners on their grilles. The logo plate on the front dropped the name of the amp, reading "Fender/Fullerton, California."

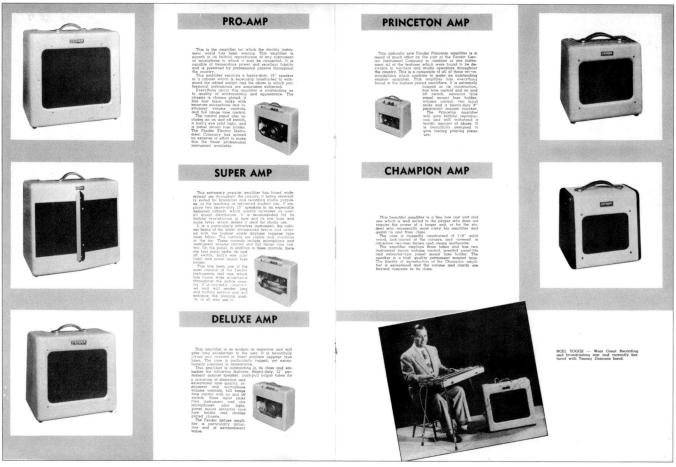

TV-front amps, 1949.

A completely different covering was applied to the short-lived Champion "800." Fender described the control panel as "A very attractive gray-green hammerloid linen finish with white markings." "A very attractive luggage type linen" was the catalog description of the green tweed that was unique to the first member of the Champion line. The control panels on early Champions were mounted to the back of the cabinet at an angle facing slightly upwards. (See page 180.)

On all models but the Champions, a new tweed with diagonal stripes began to be used c. 1949. Otherwise, there was very little changed. The Champion "800" was replaced by a new student amp called the Champion "600," featuring "attractive brown and cream two-tone leatherette." (See page 181.) Although the two-tone pattern was unique to the Champion "600," the white leatherette was experimented with on a few test models including an early Bassman and a Pro. These would date to c. 1952. All the early Bassmans from this era had ported enclosed backs and chassis mounted on the bottom of the cabinet. (See page 182.)

A new "Hi Fidelity" amp with two 12" speakers was unveiled at the summer National Association of Music Merchants (NAMM) show in 1952, featuring increased power and separate Bass and Treble controls. These controls allowed players to compensate for different acoustic environments (e.g., outdoors vs. a corner stage). Most importantly in 1952 was the growing interest in the different "sounds" available with "electric" guitars, as opposed to "electrified" ones. Having the ability to tailor one's sound through the controls of an amplifier, as well as to create a variety of sounds with the same instrument, was a fresh idea. Instruments with two pickups were still relatively new at this time. One of Fender's greatest accomplishments was its exploitation of tonal possibilities through expanded tone controls, which became increasingly sophisticated throughout the fifties and sixties, finally reaching a point of diminishing returns. In early '53 the 2x12 Twin Amp (as it was now called) began to be promoted. Following the introduction of the Twin, Fender did away with the splayed cabinet of its other dual-speaker amp, the 2x10 Super. Both the Twin and the revamped Super

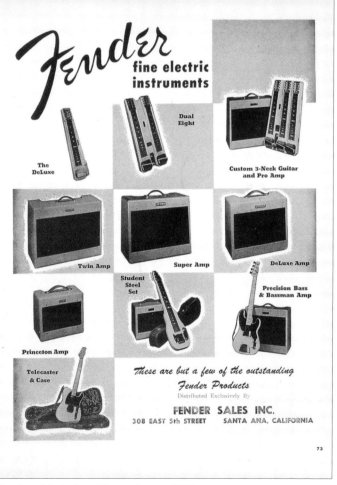

Fender ad featuring wide-panel amps, 1953.

had their speakers mounted to a flat baffle board that stretched from one side of the amp to the other. The grille was run from side to side, right up to the inside edge of the cabinet, and was no longer recessed. Two wide tweed-covered panels stretched across the top and bottom edges of the cabinet, adding a streamlined appearance and giving this cabinet style the nickname *wide-panel*. The old TV-front amps, all with single speakers, would be upgraded to this sleek new cabinet in early 1953. (See page 183.)

Narrow-panel amps, 1957. (Courtesy Hardtke Archives)

The Bassman went from one 15" to four 10" speakers and the Bandmaster went from one 15" to three 10" speakers by early and mid 1955, respectively. This required the speakers to be mounted in two rows. Again the grille had to expand, this time to the top and bottom inside edges of the cabinet. Gone were the wide panels, leaving just the frame of the cabinet covered in tweed. These are now referred to as *narrow-panel* amps, and the style would run on all the amps from 1955 until the tweed covering was retired. A new three-colored "miracle fabric" grille cloth replaced the monochromatic linen that had been in use since the late forties. A "Presence" control was added to all the amps from the Super up. (See page 184.)

Also in 1955, a Fender amp was finally available with built-in amplitude modulation, a.k.a. tremolo—available since the late forties on amps from Danelectro, Premier, and Gibson. The Tremolux was basically a souped-up Deluxe with Speed and Intensity controls. It's interesting that this popular feature was not added to the amps at the top of the Fender line for another four to five years.

Inside a narrow-panel Twin, 1958. *The X-ray shows the "big-box" cabinet, with the speakers side by side, but the little photo shows the earlier "small-box" version, with the speakers in opposite corners.*

By 1956 the 1x10 Vibrolux and Harvard, both rated at 10 watts, bridged a gap between the Deluxe and the single-power-tube Princeton. The Vibrolux supposedly introduced vibrato (pitch modulation) to Fender amps, but in reality this was just another variation on the tremolo already found in the Tremolux. In fact, *no* Fender amp has *ever* had true pitch-bending vibrato, regardless of catalog hype to the contrary. It's fitting, in an ironic way, that this confusion of terminology should come from the company that calls the vibrato arm on its guitars "tremolo."

A new steel-guitar-and-amp set for students was marketed by Fender starting in '55 and ran a number of years, under the "White" name. The amp was almost identical to the 1x8, three-tube Princeton of the time. Named in appreciation of plant manager Forrest White, these were probably made to get more inexpensive steels and amps into the market without offending authorized Fender dealers. (See page 185.)

Cosmetically, the Fender line would be relatively stable through the end of the decade, but the research and development department had a few tricks left. In '57, a midrange control was added to the Bass, Treble, and Presence controls on the Bassman, making for extremely variable tone selection—particularly when compared to the single Tone roll-off control on the Bassman until the end of '54. The crowning achievement of Fender amp designers in the fifties came in '58 with the upgrade of the Twin's power section to four 5881 power tubes. With an RMS value of between 80 and 100 watts, these amps had the ability to be very loud with minimal distortion. The difference in power and equalization between a late-fifties Twin and any competitor's finest at the time (at any time?) is astounding. Comparing it to any of the amps of the late forties, and especially the amps of the thirties, shows how far electronic components and design had progressed in a relatively short period of time. Costing one dollar less than 400, the Twin cost over 20% more than Fender's most expensive guitar of the time, the blonde-finished Stratocaster with gold hardware! (See pages 186–187.)

The company unveiled a new amp at the NAMM show in the summer of '59. The Vibrasonic offered a number of firsts for Fender, including a stock Lansing speaker, "Tolex" brand vinyl covering (in a light "pinkish" brown color), and a molded plastic handle with the company's name in raised letters. One of the most important features was bringing the top-mounted, rear-facing control panel to the front of the cabinet. Mounted along the top edge, the controls were finally available to the player who placed his amp behind him. In the thirties and forties amplifiers were usually positioned in front of or next to the performers, who were usually seated, so controls at the back made some sense. (Many early guitar amps actually had the controls at the bottom rear of the amp, probably the *least* practical place for the musician.) In the fifties, more and more musicians traded in their chairs for guitar straps. Amps were moved to the rear of the stage, creating room to move, but making it difficult to access rear-facing controls. (See page 188.)

This article is made with genuine **TOLEX**® supported vinyl fabric from the General Tire & Rubber Company as customer-styled for *Fender*®

The brown Tolex era. *Fender courts the height-impaired player, 1960.*

The Champs and their amps. *From the 1960 catalog.*

The new control panel was painted a dark brown with white lettering and was fitted with new cylindrical brown plastic knobs, smooth on the top and with ridges in the side. The dark brown panel, knobs, and handle contrasted nicely with the light brown Tolex. A flat metal logo with "Fender" in script was screwed to the grille. This amp was also the first large amp equipped with tremolo. Later that year another new model, the 4x10 Concert, was released with the same appointments except for Jensen speakers. All the amp fronts at this time were covered in the tweed-era two-tone brown with red threads.

In early 1960 all the amps in the "Professional" series were upgraded to the new style introduced by the Vibrasonic. These included the 2x10 Super, 3x10 Bandmaster, 1x15 Pro, and 2x12 Twin. All the light brown Tolex amps were equipped with tremolo and had the Volume knob placed in the middle of the controls, reading left to right from the inputs as follows: Input, Bass, Treble, Volume, Speed, Intensity, and Presence. Smaller amps—the 1x6 Champ, 1x8 Princeton, 1x10 Harvard, 1x10 Vibrolux, 1x12 Deluxe, and 1x12 Tremolux—remained unchanged, as did the 4x10 Bassman, which would never receive tremolo. (It's interesting that today this amp is considered by many to be the ultimate guitar amp.) Plans were probably already being made for the Bassman, whose days as a combo were numbered.

Later in the year, the Volume control on the "Professional" amps was switched with the Bass control, so that left to right from the inputs the panel would read: Input, Volume, Treble, Bass, Speed, Intensity, and Presence. (See page 189.)

A photo in the 1960 catalog showed popular recording group the Champs in the studio, outfitted with Fender guitars and amps. Most interesting is the stack of light brown Tolex cabinets with the new logo. One of these was a prototype piggyback amp. The formal

PART I: HISTORY—*FENDER AMP LINEAGE*

announcement came in December of '60. The Showman was essentially the chassis of a light brown Tolex Twin in a small box, with a separate box housing a JBL speaker. The speaker cabinet had an elaborate porting system. The cabinets were covered in Tolex, but in a new color: white. The knobs were also white, and a new color—maroon—was picked for the grille cloth. This combination of cosmetic features is classic and has been revived by the company for some of its top-of-the-line Custom Amp Shop models. (See page 190.)

In early '61 the Twin switched to the color scheme of white Tolex and maroon grille cloth, as did three previously available combos that were converted to piggybacks: The 4x10 Bassman, which was still tweed in 1960, became a 1x12 piggyback; the 3x10 light brown Tolex Bandmaster was also changed to a 1x12 cabinet; and a few months later the 1x12 tweed Tremolux became a 1x10 piggyback. The new maroon grille cloth soon replaced the tweed-era cloth on the large combos, which were still covered in the light brown Tolex. The top front edge of the cabinets was sharp at this time. A new dark brown Tolex soon replaced the earlier shade, and the top edge became rounded. (See page 191.)

Piggyback amps. *From the 1961 catalog.*

The smaller amps (excluding the Champ) had their tweed covering replaced with the new dark brown Tolex, but were fitted with wheat-colored grille cloth. The '61 catalog shows the baffle boards on these amps screwed in from the front, as on later Reverb Units. The early Reverb Unit shown in the catalog had the front panel covered in Tolex instead of grille cloth. In '62, the wheat grille replaced the maroon on the large combos. (See page 192.)

The piggybacks were still shown with maroon grilles, but these, too, would change to wheat. Two Bassmans, both dated October '62, one with a maroon grille and one with wheat, show that the change occurred late in the year. A Twin similar to the one pictured on page 193 was found to have a date of December '63.

In early '63 more changes occurred in both the circuitry and the cosmetics. In the February issue of *Fender Facts* (a public-relations newsletter sent to dealers and owners of Fender equipment), the new Vibroverb was announced, featuring built-in reverb for the first time. The amp pictured had the brown leather handle standard on the smaller combos of the time. Production models came with a new handle of metal-reinforced flat plastic, with metal caps at the ends. (See page 193.) These would soon become standard on all the Fender amps and were basically unbreakable, unlike the brown molded handles, which snapped in cold weather, and the leather ones, which simply wore out (usually while the amp was being picked up or carried, often causing it to fall and break). The new handles were all black, which looked fine on the white Tolex amps, but was rather unsightly on the brown. (The author searched for a brown Vibroverb from 1979 to 1983 before finding one, and then traded it away because he couldn't stand looking at the handle.) Why they didn't order these in brown to match their amps of the time is puzzling, unless you consider their plans to release the new black Tolex models at the NAMM show that summer.

The '63 catalog showed the line in a very transitional stage. On the cover was the Twin Reverb, one of a number of new combo amps with built-in reverb and covered in a smooth black Tolex. The knobs were no longer cylindrical, and the numbers were no longer painted on the control panel. Black-skirted knobs with silver caps and numbers 1 through 10 became standard on all the amps (with a few exceptions), as did the black control panel—which gave these amps the nickname *blackface*. The gray grille cloth pictured was only used on prototypes and was replaced with the "silver sparkle" grille cloth that would become standard on all Fender amps by '65. The old-style flat logo was already replaced with a new, "raised" logo of chrome on black. (See page 194.)

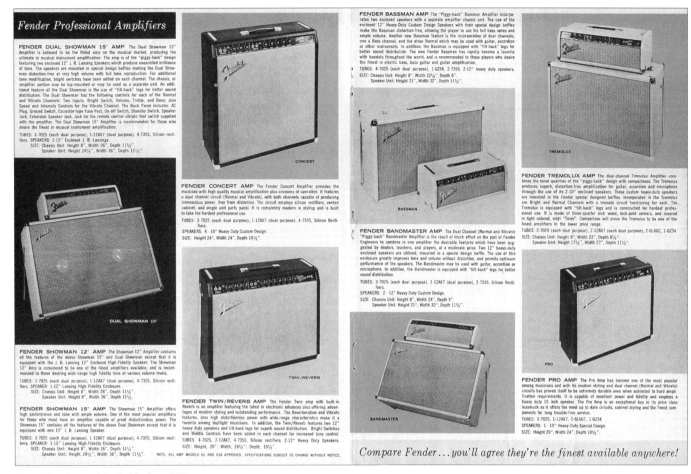

Transition from white to black, c. '63.

The piggybacks with rough white Tolex, wheat grilles, and flat logos were made for close to a year, although most were equipped with white knobs and the early-sixties electronics. The very last ones had the new circuitry and black, numbered knobs. (See page 195.)

The covering on the piggybacks was soon changed to a smooth white Tolex. The wheat grille cloth had sparkly gold woven into it starting at the end of '63, and this variety appeared on the last of the white piggybacks. (See page 196.) The Bassman amps did not receive the new black knobs, possibly to use up old parts, but were some of the best-looking amps ever made. The combination of smooth white Tolex, white knobs, black control panel, and gold sparkle grille, with black-and-chrome handle and logo, appear to have been used only on the '63–'64 Bassmans and Reverb units—somewhat out of the logical progression of features, but definitely production models. (See page 197.)

FENDER AMPS: *THE FIRST FIFTY YEARS*

From the middle of '63 to late '67 Fender made thousands of blackface amps, most of which are still working fine today. Certainly not rare, they are valued for their design, both electronically and aesthetically. To many musicians, they represent the pinnacle of Leo Fender's efforts. (See pages 198–199.)

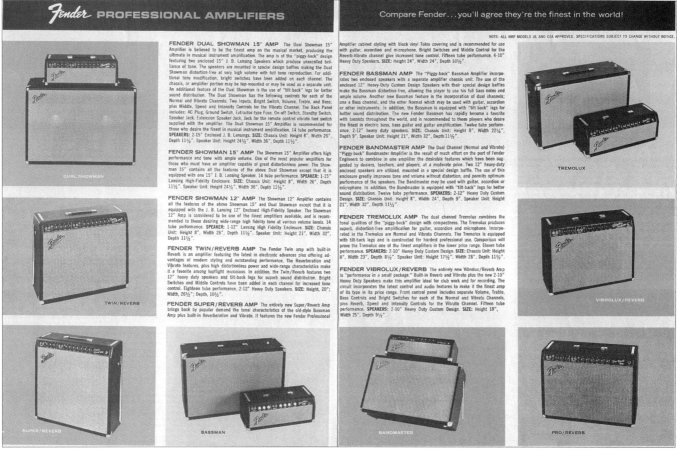

Blackface, '65.

In early '63 the Fender line included 1x12, 2x10, 1x15, and 4x10 combo amps with brown Tolex, two 6L6s, and no reverb. In '65 the two-6L6 combos were 2x10, 2x12, and 4x10, all with reverb, but with names not linked up to the '63 models. Things would make more sense, and amps would have more direct ties to their past, if Fender would have called the brown Vibroverb the Super Reverb. Then we'd have a 4x10 Concert Reverb instead of the Super Reverb, a 1x15 Pro Reverb instead of the black Vibroverb, a new improved Vibrolux Reverb with an extra 12" instead of the Pro Reverb, and a 2x10 Super Reverb instead of the Vibrolux Reverb. Unfortunately, it didn't happen that way.

Tweed had survived into 1964, being used to cover the bottom-of-the-line Champ. These amps had basically been ignored by the R&D team for years, still having a rear-facing, top-mounted control panel and a pointer knob for the volume control. A transitional version stayed basically the same, but with black Tolex replacing the tweed, and a black plastic handle replacing the old leather one. This short-lived variant was replaced in late '64 by a version with a front control panel that included Volume, Bass, and Treble controls, and the Vibro Champ, which had two more controls, one more tube, and cost 10 more dollars. A list of other short-lived blackface models includes the 1x15 Pro (replaced by the 2x12 Pro Reverb in early '65), the 1x12 Vibrolux (replaced by the 2x10 Vibrolux Reverb in mid '64), the fleeting 1x15 Vibroverb (summer of '63 through fall of '64), and the 4x10 Concert (discontinued in early '65). Transitional Bassman and Princeton amps c. '63–'64 featured black Tolex and white knobs (a very sharp look!). The matching Reverb Unit was unchanged through the end of '66, when it was discontinued. (See page 200.)

The words "Fender Electric Instruments," or any of the various abbreviations "Fender Electric Inst.," "Fender Elect. Inst.," etc., appeared underneath the pilot light on all the amps with front control panels from '59 to the fall of '65 (in a number of different type styles). These were replaced with the words "Fender Musical Instruments," which to many denote CBS-era amps. Otherwise, the amps were identical, built in the same factory, using the same parts. In reality, any amps (or guitars) with a date code after December '64 are from the CBS era. Blackface amps were available at least through the last month of '67 with no obvious change in quality, the CBS-era models having been available for twice the length of time as their pre-CBS counterparts, and selling in greater numbers. This should help dispel the myth that "pre-CBS = good, CBS = bad." (Listen with your ears!)

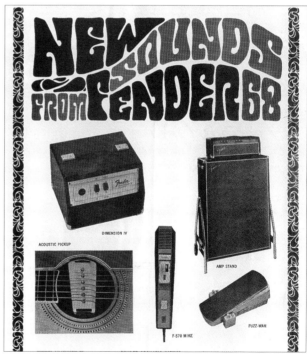

Miscellaneous products from the CBS era.

A line of amps that didn't help CBS's cause was released in July '66. Two words that instill fear in the hearts of purists, "Solid-State," described a whole new method of amplifying music. The power-vs.-price specs that were employed to push the cheaper-to-make transistor amps described the response of amplifying a sine wave, and on paper seemed like a good thing. Unfortunately, a lot goes on in a tube that specs don't describe. The solid-state series introduced a new grille cloth with blue woven into the material. Aluminum-colored plastic trim outlined the grille,

FENDER AMPS: *THE FIRST FIFTY YEARS*

First-series solid-state amps, '66–'71. *Dark days for Fender.*

and metal knobs on a silver face plate were a few of the other new features. Public response to the solid-state line was slow, but by the summer of '68 the company offered solid-state versions of seven of the 10 tube amps in the line. (See page 201.)

By the fall of 1969, a totally new solid-state line replaced all but the original Bassman and the Super Showman. The new "Zodiac" amps featured cosmetics never seen before or after their short run. Alligator covering, new grille cloth, and a vented control panel were a few of the unique features of the trend-riding line, which was retired in '71. (See page 202.)

By 1968, the tube amps began to change cosmetically, receiving blue grille cloth and "aluminum" trim with silver control panels (*silverface*) and blue block lettering. Otherwise, they were about the same as earlier versions. The Bronco amp was fitted with new metal-skirted knobs and red lettering. For a short time the raised logo lost its "tail," but the company quickly returned to the earlier version. Minor changes in the circuitry began to happen, but a knowledgeable repairman can make these later versions sound great with a minimal amount of work. (See page 203.)

Silverface with aluminum trim, 1969.

Effects of the seventies ('76). *From Fender's wacky 1976 catalog.*

The aluminum trim was dropped in '69, and another tail-less logo (with an "®" symbol) began to be used in the mid seventies; otherwise the amps didn't change much cosmetically. Electronically, they were "upgraded" with a master volume on the large amps in the early seventies and a pull-knob "Boost" control on the volume pot in the late seventies. The logo added the words "Made in U.S.A." These were the biggest changes in a remarkably stable line, running into 1981. (See page 204.)

The CBS R&D department was far from dormant in these years, but chose to do its experimenting on new models, like the solid-state Super Showman, leaving a dependable tube line available to the public. (CBS also experimented with prices—exactly once. After boosting them substantially across the line at the beginning of 1970, they were forced by public outcry to bring them back down almost immediately.) Models like the Bantam and Musicmaster Bass amps and the Quad and Super Six Reverbs were appointed like the rest of the line, while the high-powered 400 PS, 300 PS, and Super Twin were dressed up with new black control panels, black grille cloth, and white trim. (See page 205.)

Super Twin Reverb ('77–'80), '79. *These amps were incredibly loud—Ted Nugent only needed six.*

The tried and true line of silverface tube amps, basically unchanged for over 10 years, received a face-lift in '82, going back to the black control panel and silver grille cloth of a previous generation. The white lettering, which now ran straight across instead of at an angle, was one of a number of minor differences separating the '82 blackface amps from mid-sixties versions. These were from the time of the first reissue guitars, showing Fender's awareness of their past. (See page 206.)

Blackface, second version ('82). *Kids, don't try this at home!*

Later that year, a new line of amps designed by Paul Rivera replaced the traditional tremolo-equipped models. The new "II" series amps looked similar, but had high-gain circuitry with pull-knob boosts and channel switching. (See page 207.) A "special edition" series was released, featuring oak cabinets with a natural finish, brown control panels, light brown grille, and heavy-duty brown leather handles. The Super Champ, Princeton Reverb II, and Concert were the only amps available with this package. These would be the last amps of the CBS era.

For a short time just before CBS sold the company, a small line of solid-state amps was again offered. The Harvard amps of late '81 were joined in '83 by a half-dozen other models, all with similar features, including a small plastic

"II" series amps ('83), '82–'86.

The truth is... ('84). Quality and innovation (including another try at solid-state design) couldn't dissuade CBS from selling Fender in early '85.

plate with the name—e.g., Montreux, London, etc. The knobs on these amps were similar to the standard skirted ones, but had flat sides and lacked the chrome caps. In '84 Fender began to import the Sidekick series from Japan. The logos on the imports don't say "Made in U.S.A."

Sidekick amps ('87), '84–'94. This line of small solid-state units, initially imported from the Far East, kept Fender in the amp market when the company changed hands.

Following the sale of the company in early 1985, the new management had no factory for making amps; it relied on the imported Sidekick models and a supply of leftover "II" series amps, which lasted into late '86. The first tube amps from the new Oregon factory were the Champ 12, "The Twin," and the Dual Showman, released in late '86. The amps were similar to the models of late in the CBS era, with black Tolex, handle, control panel, and knobs. The flat-sided knobs differed from the earlier ones, lacking printed numbers, which were instead stenciled on the control panel. On the first Champ 12s, the grille cloth was silver. The knobs were soon changed from black to red, and the grille cloth to a dark gray/black. "The Twin" and the Dual Showman, released with red knobs, were essentially the same 100-watt amp—"The Twin" a 2x12 combo, and the Dual Showman a separate head with 4x12 cabinets. The Champ 12, aimed at the home user, featured a headphone jack and tape inputs. The amps were available in red, white, gray, and "snakeskin" custom coverings, as well as black. A club-sized amp, the Super 60, was announced in January of '88 and released later that year, having the same appointments as the Champ 12, Dual Showman, and "The Twin." (See page 208.)

In '89 a new series of American-made solid-state amps ranging in power from 50 to 160 watts joined the small line of 12-, 60-, and 100-watt tube amps. The cosmetics of these solid-state amps (red knobs, gray grille cloth, etc.) matched the tube amps.

The reissue guitars of the early eighties had rejuvenated the Fender company in the public's eyes, and in 1990 the new Fender Musical Instruments Corp. followed suit on the amp front, this time with reissues of two old models that had become "collector's items" for their usefulness as well as for their "cool factor." The '59 Bassman and '63 Vibroverb reissues were soon the rage amongst bar bands and professionals alike. For a time it seemed as if every blues player was using the Bassman reissue. The original Vibroverb is one of the rarest Fender amplifiers. The reissue made the classic combo available to all at a third the price of an original. Tremolo was again available on a Fender amp (for the first time since 1982), making the gods happy! In 1991 the '65 Twin Reverb Reissue was added, and in 1994 a blackface '65 Deluxe Reverb Reissue joined the line. (See page 209.)

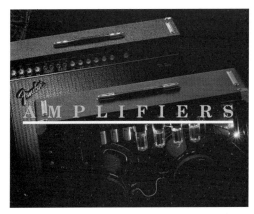

"Red-Knob" series ('89), '86–'94. *The new Fender Musical Instruments Corp. steps out.*

Mini amp ('89). *Admit it: You think they're cool.*

Amps for all players (c. '90).

Reissue, Professional Tube, and Custom Amp Shop series ('93).

The nineties saw Fender make a variety of fashion statements with their new releases. A whole new look was given to a new line of high-gain solid-state amps. The "R.A.D.-H.O.T.-J.A.M." series of small combos, and the M-80 series of high-power amps, were covered in a light gray—and later, black—carpet material with black plastic corners. This was a drastic (and short-lived) experiment. The Tweed series reintroduced—you guessed it—tweed to the regular line. (See page 210.) At the top of the Fender line today is the Custom Amp Shop series, inaugurated in '93. (See page 211.) These hand-wired amps promise an intriguing array of coverings, ranging from white Tolex (with maroon grille cloth, of course) to "Sea Foam Green Lizard." Who knows what the future holds?

PART II: THE AMPS
TUBE AMPS

DELUXE
1945–1985
1994–Present

Deluxe
Model 26
Deluxe Reverb
Deluxe Reverb II

If the Fender amp line has an "Old Standby," it is the Deluxe. The first model, also referred to as the "Model 26," goes back to 1946 and has direct ties to the 1x10 K&F model of 1945. Although "Model 26" was printed in small letters on the control panel of these early models, the company referred to them as the "Deluxe" model, a name still used by Fender. A single 10" speaker and five tubes may not seem like much today, but in '46 it satisfied most playing situations. The cabinets were tastefully constructed from a choice of hardwoods (maple, walnut, and mahogany) and fitted with a variety of colorful grille cloths (red, blue, and gold). The back panel of each amp was covered in the same colored cloth as the front. Three (occasionally two) vertical metal slats across the front protected the speaker and added a bit of flash to the amp when hit by lights. A simple wooden handle was screwed into the top and in some examples actually held the chassis in place. At least two other methods of mounting the chassis were used, with bolts through the top or the sides.

"Woodie" Deluxe ('46), '46–'48. *This early model fitted with two slats instead of the more common three. These are a different width than later models', but of the same metal.*

Somewhat fancier than its little sister Princeton, the Deluxe had a stenciled control panel with the impressive Fender lightning-bolt logo, three input jacks, Mic and Instrument Volumes that went to 12, and a combination On-Off switch/Tone control, all with pointer knobs. The tone roll-off function worked backwards on the early Fenders, with fully clockwise yielding minimum treble. The high-impedance Mic input went directly to the grid of the cathode-biased preamp tube. The Instrument inputs were somewhat padded down by two 56kΩ resistors that isolated the two jacks from each other. Tube configurations changed regularly as the new company experimented and upgraded the circuit design. An early version (e.g., serial no. 219) used a 6SN7 twin triode tube for the preamp, with one half for the Mic input and the other half for the Instrument Volume. A second 6SN7 was used for the *paraphase* inverter, with two

Fender's Model 26 circuit was actually quite versatile for the time—separate Instrument and Mic Volume controls, with a master Tone knob.

cathode-biased 6V6s for the power section, and a 5Y3 rectifier. The glass 6SN7 phase inverter tube was replaced (e.g., serial no. 434) with a metal-cased 6N7, also found on the Professional and Dual Professional models.

These early Deluxes lacked circuit boards, instead having their components connected directly to each other (without wire) wherever possible, with ground connections soldered directly to the inside surface of the chassis. The tubes were mounted vertically, hanging upside down from the bottom of the chassis. The addition of a circuit board for making the electrical connections (around serial no. 500) coincided with a new shape for the chassis. On these models, the tubes were mounted horizontally. The 6SN7 preamp tube was replaced (e.g., serial no. 806) with a metal-cased 6SC7, and the possibility of other configurations certainly exists. The early models used field-coil speakers, which were soon replaced with permanent-magnet Jensens. The output transformer was mounted to the chassis on the field-coil versions, while on the permanent-magnet versions it was mounted to the frame of the speaker, a common method for many amp manufacturers during this period.

What a difference a year makes! *Early Deluxe (on right in both photos), no circuit board, c. '46. Later version (on left), with circuit board, c. '47. Note tube orientation.*

Following the release of the tweed-covered Dual Professional, c. 1947, the wooden Deluxe followed suit (by the summer of '48), also being upgraded in a number of other areas. The new "vertical"-tweed-covered *TV-front* cabinet housed a 12" speaker behind dark brown mohair grille cloth. A new chassis, chrome-plated with the numbers and new logo stenciled on in white, included two new additions: a pilot light and a fuse holder. The 6N7 phase inverter was replaced with another 6SC7, and the preamp tubes changed to grid resistor bias for these amps, which would be the basis for the tweed Deluxe amps that followed (e.g., model 5A3). And the Deluxe became a *very* popular amp. "As modern as tomorrow."

TV-front Deluxe ('48), '48–'53. *First tweed Deluxe.*

The combination of vertical tweed and mohair grille was replaced with "diagonal" tweed and brown linen grille cloth c. 1949, but little was changed the next few years. The *wide-panel* cabinet replaced the TV-front c. 1953, but otherwise the model was not much different from the '48 model (e.g., wide-panel model 5B3, serial no. 6880, with pot codes of the fifty-second week

Wide-panel Deluxe ('54), '53–'55. *Transitional. Started in '53 with metal tubes from TV-front Deluxe era—ended in '55 with glass 12AX7s, etc., a circuit that would last until 1960.*

of '52). Tubes were still two 6SC7s, with glass-bottle 6V6GTs and 5Y3GT substituted for the old metal versions. A separate On/Off switch and a spare speaker output jack were obvious additions; a negative-feedback loop was less obvious, but certainly affected the sound more (model 5C3).

This upgrade was deleted from the following model (5D3), which replaced the grid-resistor-biased 6SC7 preamp tube with a cathode-biased 12AY7. A 12AX7 *self-balancing paraphase* inverter replaced the other 6SC7, leaving the metal tubes gone forever.

Narrow-panel Deluxe ('55), '55–'60. *A classic.*

In '55 the *narrow-panel* style Deluxe was introduced, featuring genuine grille cloth, the new ground switch, and four inputs (model 5E3). Each of the four was connected to a 68kΩ resistor, an input circuit still used by Fender for many of its amps. There was no difference between the Mic and the Instrument channels. The phase inverter circuit was changed to a *split load*, requiring only half of the 12AX7, leaving the other half as a second preamp stage, common to both channels. Like all preceding versions, the narrow-panel Deluxe used cathode bias for the two 6V6 power tubes.

The first narrow-panel Deluxes had a new script logo and measured 16½"x18"x8¾". These dimensions were soon enlarged to 16¾"x20"x9½", around the release of the Vibrolux in '56. These two amps would share cabinets until 1961 and the change to Tolex. The nameplate added the words "Deluxe" c. 1957 and "Fullerton California" for the last models, but very little changed in the circuitry (e.g., model 5E3 from 1960).

Brown Tolex Deluxe ('61), '61–'63. *Very early model with baffle board mounting from front; production models would have Normal channel on left side, and Bright next to Tremolo controls.*

A total overhaul of the cosmetics took place in time for the '61 catalog, with the control panel moving to the front, the tweed being replaced with brown Tolex, and the addition of tremolo. The new dark brown control panel was fitted with brown knobs and a new layout: Volume and Tone in the Bright channel and Volume, Tone, Speed, and Intensity in the Normal channel. Actually, the tremolo worked on both channels, by varying the new *fixed bias* voltage on the power tube grids. Only the first brown Tolex Deluxes were set up in this order, with standard models having a white line separating the tremolo controls from the others.

The channels were reversed on these models, with Normal on the left and Bright next to the tremolo controls (model 6G3). Speaker and extension jacks were mounted on the rear panel, as was a jack for the tremolo pedal and the On/Off switch. The 12AY7 preamp tube was replaced with a 7025, and another 12AX7 was added, with half for the tremolo circuit oscillator. The new *long-tailed* phase inverter, first seen on the big amps of the late fifties, used both sides of the other 12AX7. Although equipped with a number of new features, the brown Tolex Deluxe still had the two inputs tied together following the volume controls, causing interaction between the two channels.

The model pictured in the '61 catalog had wheat-colored grille cloth, with the baffle board secured to the new cabinet (17½"x20"x9½") by four screws applied from the front (in a similar fashion to later Reverb Units). Other small amps that were introduced at this time were shown assembled in this manner, but production models were screwed in from the back, as seen in the 1962 catalog.

The Deluxe had a big year in 1963. Black handle, control panel, knobs, and Tolex were highlighted by the silver sparkle grille, in classic blackface style. Like all the small combos, the Deluxe lacked the new raised logo. The cabinet was enlarged (17½"x22"x9½") to accommodate the wider head (model AA763), which now included separate Bass and Treble controls for each channel. Two 7025s, one for each channel, allowed the controls to be placed in the circuit between the two stages of each tube. This eliminated any interaction between the controls of the two channels as on earlier Deluxes. The common preamp stage of the previous model was deemed unnecessary, freeing up half of a 12AX7. This was added to the tremolo circuit, which was a brand-new design common to the blackface amps. The Vibrato channel preamp signal (slightly brighter than the Normal) connected to the tremolo Intensity control, which in turn connected to a photoresistor. This device varied its resistance in time with the tube oscillator, allowing a path to ground (grounding out) for the preamp signal. The Normal channel was now unaffected by the tremolo. A 12AT7 replaced the 12AX7 in the phase inverter and went on to be

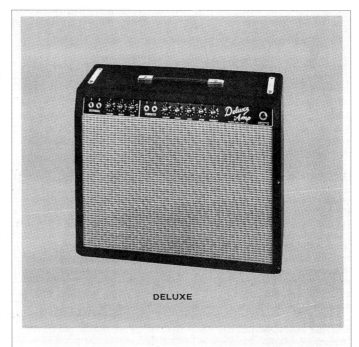

FENDER DELUXE AMP WITH VIBRATO The fine styling and performance of the Deluxe Amp is the same as the above Deluxe/Reverb except that Reverb is not included. Controls are identical with the exception of the Reverb Control and the Reverb Foot Pedal. **SIZE:** Height 17½", Depth 9½", Width 22". 1-12" Heavy Duty Speaker. Seven tubes, four dual purpose.

Blackface Deluxe ('65), '63–'66. *Short-lived model, without reverb.*

Fender's standard choice of tube for this function. The new circuitry included a courtesy AC outlet and a 10kΩ Bias Adjust pot for the output tubes (still two 6V6s, with a GZ34 rectifier).

The *Down Beat* insert of September '63 showed a Deluxe with the solid gray cloth common to the first blackface models. This flyer introduced two new products that would soon become industry standards: the Deluxe Reverb and the Twin Reverb. The Deluxe Reverb (model AA763) was also fitted with the new blackface, black Tolex style cabinet with solid gray grille (17½"x24"x9½"). Production models, as seen in the '63–'64 full-line catalog, were dressed in the standard silver sparkle grille cloth. The Reverb knob and the rear-panel jack for the second half of the new dual foot switch (Vibrato/Reverb) were obvious differences between the Deluxe and the Deluxe Reverb, which also had two more tubes. This tube lineup would be used on most of the two-channel reverb amps that followed; they are, looking at the backside of the amp, left to right, rectifier (GZ34), power tubes (two 6V6), phase inverter (12AT7), tremolo (12AX7), reverb return/Vibrato channel third preamp stage (7025), reverb driver (12AT7), Vibrato channel preamp (7025), and Normal channel preamp (7025).

Deluxe Reverb ('70), '63–'82. Silverface Deluxe Reverb ('68–'81), nearly identical to blackface ('63–'67). The first silverface amps were fitted with aluminum trim. Went back to blackface for last two years.

Little change was made in these amps for the next four years, with "Fender Musical Instruments" replacing "Fender Elect. Inst." in mid '65 and the addition of the raised logo c. 1966. The Deluxe was phased out in the fall of '66, but the Deluxe Reverb would remain in the line, with only minor changes, until 1982. These amps are all highly prized for their portability and their ability to be cranked in small clubs. The price of a Deluxe Reverb on the used market is about equal to that of comparable era Twin Reverbs, which originally cost twice as much and were made in smaller quantities—a perfect example of supply and demand.

The 20-watt blackface style was replaced by the 20-watt silverface model with blue sparkle grille and aluminum trim c. 1968. Electronically, the amplifier would stay basically unchanged except for the addition of two 1200pF capacitors connecting the power tube grids to ground (model AB868). The bias section did not change on the Deluxe Reverb, as it did on the larger amps, making silverface models a very smart buy these days. The aluminum trim was discontinued around the end of '69, a three-position Ground Switch was added in '70, the raised logo lost its tail in the mid seventies, and a pull-knob Volume/Boost pot became standard in the late seventies. A Line Out jack completed the electrical changes. The blackface, silver sparkle grille option was offered in 1980 and became standard a year later.

An entirely new amp appeared in '82—the Deluxe Reverb II. The "outdated" tremolo circuit was replaced with distortion and channel switching. Preamp Output and Power Amp Input jacks were added to the rear panel, as was a Hum Balance control. The controls for the clean channel were Volume with pull-knob for Bright, Treble, and Bass (the reverb controlled both channels). The high-gain channel controls were Volume with pull-knob for Channel Select, Gain, Master, Treble, Mid with pull-knob for Boost, Bass, Reverb, and Presence. A 7025 phase inverter, two 6V6GTA power tubes, and a solid-state rectifier made up the power section. The traditional use of a 12AT7 and half of a 12AX7 for the reverb circuit was not unusual for Fender, but the preamp section was radically new. Six preamp stages (three 7025s) were used between the two channels. The On/Off switch was moved to the front panel, joining the two input jacks and a red pilot light. The Deluxe II was discontinued at the time of CBS' disposal of manufacturing, with existing stock sold by the new company into late '86.

Deluxe Reverb II ('83), '82–'86. *Hot-rod version designed by Paul Rivera and Ed Jahns.*

Following a short period surrounding 1987 when the name was not used, a solid-state model—the Deluxe 185—was released in '88. This had absolutely nothing in common with the tube Deluxes of the past, nor did the solid-state Deluxe 112 of '92. Finally this disservice to the amp's good name was remedied with the release of the '65 Deluxe Reverb Reissue of '94. If the Reissue does what the other Reissue amps have done, it should be a huge success; the Deluxe Reverb is one of the all-time great amps, period.

Deluxe Reverb Reissue ('93), '93–. *Reissue of highly sought-after blackface model. A limited-edition version was also issued, in historically incorrect but tasteful white Tolex with gold grille cloth.*

PRINCETON
1945–1985

Princeton
Princeton Reverb
Princeton Reverb II

The Princeton name goes all the way back to the original trio of Fender amps from 1946. Though it had no model name, the 1x8 K&F was the obvious predecessor to the wooden-boxed 1x8 Princeton. These two entry-level amps used the most basic circuit possible—three tubes (6SL7, 6V6, 5Y3) and no controls! The Princeton steel guitar had a volume pot built into it, so why put one on the amp? An On/Off switch? Unplug it from the wall when you're done!

"Woodie" Princeton ('46), '46–'48. *Three tubes, no Volume or Tone control, removable internal fuse. Inside of chassis similar to K&F 1x8.*

The amp was equipped with not one, but two input jacks, to accommodate student/teacher sessions. (The guitar, by the way, was "jackless," the cable being hard-wired to the volume control.) On both these amps, the jacks were secured to the backside of the chassis, *sans* control panel, logo, numbers,…anything. The K&Fs had metal grilles and gray crinkle finish; the wooden Princetons were offered in the same woods and grilles as the bigger models and featured the three chrome stripes across the speaker opening. Both models used Jensen 8" field-coil speakers, with the Princeton probably changing to a permanent-magnet speaker c. 1947. Components were tied directly together without connecting wires where possible, due to the lack of a circuit board. The grounds were soldered directly to the chassis. A number of variations in tube configurations (6SJ7 preamp), alignment, and component layout show the transitional stage Fender was in at the time.

TV-front Princeton ('48), '48–'53. *New model with Volume and Tone controls. Vertical tweed covering on first version; diagonal tweed c. '49.*

With the introduction of the Champion line of student amps and steel guitars, the Princeton was upgraded, leaving the factory with volume *and* tone controls mounted to the control panel. An On/Off switch was part of the Tone control, clicking off when turned all the way counterclockwise. A fuse holder, a jeweled pilot light, and pointer knobs completed the package. The 1948 model featured a new

PART II: THE AMPS—*PRINCETON*

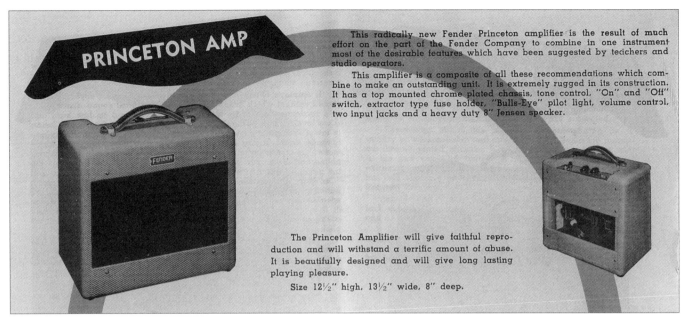

Wide-panel Princeton ('54), '53–'55.

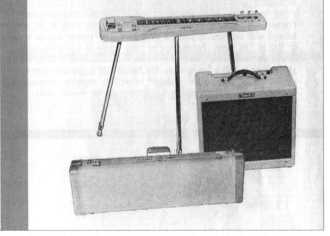

Narrow-panel Princeton ('58), '55–'61. *Studio Deluxe Set included Princeton amp; also available solo. Early narrow-panel model had smaller cabinet.*

"vertical"-tweed-covered *TV-front* box with the chassis mounted to the top, with the chrome control panel facing up. Grille cloth was a fuzzy mohair material. By the end of the forties, diagonal tweed with brown linen cloth was standard; otherwise the amp stayed basically the same until '53, when it changed to the *wide-panel* cabinet. The metal-cased 6SL7 was replaced c. 1954 with a glass 12AX7. The controls were moved from directly off the grid of the power tube to between the two stages of the 12AX7 (model 5D2). The wide-panel Princetons measured 12½"x13½"x8", as did the *narrow-panel* version of '55 with the new grille cloth. A choke was added to the filter section of the power supply, and a negative-feedback loop was added to the circuit (model 5E2).

49

Brown Tolex Princeton ('62), '61–'63. *Much more amp for the same price: 10" speaker, two 6V6s, and tremolo.*

Black Tolex Princeton with white knobs ('64), '63–'64. *Blackface Princeton Reverb? I don't think so! The black-knobbed Princeton and Princeton Reverb would finally show up in late '64.*

The choke, an expensive addition on the inexpensive amp, was soon removed (model 5F2-A). Fender rated these single-ended Class-A wonders at 4½ watts. The '56 model was enlarged to a 16½"x18"x8¾" box, also used for the Harvard. The Champ got the old Princeton box.

The 1960 catalog description read, "This radically new Fender Amplifier…," the same text the company had been using since 1948! Nineteen sixty-one brought a complete change to the model, which was basically rebuilt from the ground up. A brown Tolex cabinet (16½"x19"x9") with a wheat grille surrounded a 10" speaker and a brand-new chassis (model 6G2). A front-facing brown control panel held pots with brown knobs for Volume, Tone, Speed, and Intensity. The latter two controls operated the newly added tremolo circuit, which used only half of a 12AX7 and worked directly on the power tube's new fixed bias section, similar to that of the tweed Vibrolux and Tremolux models and very unlike the "Harmonic Vibrato" section of the new "Professional" series. Another big change in the circuit was the upgrade to two 7025 preamp and two 6V6 power tubes, yielding approximately 12 watts. A ground switch was added on the rear panel at this time. This model was more akin to the single-10", two-6V6, tweed Vibrolux than the previous Princeton. Only the leather handle was left.

An interesting transitional model was pictured in the '63–'64 catalog. The box and chassis were unchanged but were dressed up in the new black Tolex, with a black plastic handle. The grille cloth shown was the early prototype solid gray seen on a number of the amps at the time, but the new black-skirted knobs with numbers were placed over the old numbered panel for this photo shoot. Production models as pictured in the '64 catalog with standard silver sparkle grille cloth were fitted with the white knobs previously seen on the big boys. What a looker this amp is! Coinciding with the release of the Princeton Reverb c. late '64, the Princeton changed over to the black, numbered knobs, gaining separate Treble and Bass controls (model AA964). A GZ34 replaced the

5Y3, adding considerably higher plate voltages (the transformers were unchanged), but otherwise these two amps are much more similar to their predecessors than the larger blackface amps. The cabinet was down-sized to 15⅞"x15⅞"x8½".

The Princeton Reverb (model AA1164) was essentially the same amp as the Princeton but with Reverb and would go on to become a favorite of studio musicians, being the smallest Fender with tremolo and reverb. The model was described in the '64 full-line catalog, but the picture was of a white-knobbed Princeton with the name covered up by a guitar. The *Down Beat* insert from '64 and *Fender Facts* #7 (August '64) also described the Princeton Reverb but did not show a photo. It finally appeared in the '65 catalog. Little was changed on these two amps over the years other than the regular cosmetic "upgrades." They received raised logos in '66, silverface with blue grille cloth and aluminum trim for '68 and '69, and the standard silverface style all through the seventies. A three-position ground switch was added to the back panel in '70, but otherwise the circuit (AB1270) was *identical* to the blackface models, making these amps a real bargain today. The tail was dropped from the logo in the mid seventies, and a Boost pull-knob pot was added shortly before the Princeton was dropped in '79. The Princeton Reverb held on a few more years. Blackface with silver grille became an option in 1980 and finally replaced the dozen-year-old silverface style.

Blackface Princeton Reverb ('67), '64–'67. *Idyllic setting, but would you plug in?*

Silverface Princeton and Princeton Reverb ('72), '68–'81. *Similar to blackface.... Models from '68 to '69 fitted with aluminum trim. The Princeton Amp was retired in '79; last Princeton Reverbs returned to blackface c. '81.*

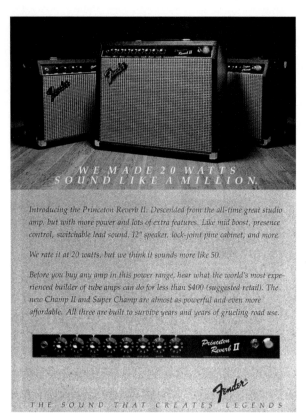

Princeton Reverb II ('82), '82–'86.

This last variant was short-lived and was replaced by the Princeton Reverb II in '82. Part of the new series that exchanged tremolo for distortion, the Princeton Reverb II was more powerful (20 watts) and had a bigger speaker (12") and cabinet (20⅝"x16⅞"x10⅞" replacing 19¾"x16¼"x9½"). The front-panel layout was one Input, Volume with pull-knob for Channel Select, Treble with pull-knob for Bright, Mid with pull-knob for Boost, Bass, Reverb, Lead Level, Master, and Presence. All of these were made by CBS, although they continued to be sold by FMIC into late '86. The name, which went all the way back to the beginning of the Fender amp line, was demoted for use on the Princeton Chorus ('86) and Princeton 112 ('92) solid-state combos....

PRO
1946–1982

Professional
Pro
Pro Reverb

For the select few that needed a little more than what the 1946 Deluxe offered, Fender had a step-up model that was their genuine best effort: the Professional. The wooden cabinet was a larger version of the hardwood Deluxe and Princeton models, with metal protectors mounted across solid red, blue, or gold grilles. The components used were the best available and at the time were really more than most players needed. The Jensen 15" field-coil speaker was about as good as it got and cost more than twice the price—and handled twice the power—of the Deluxe's 10". The two metal 6L6 power tubes were capable of putting out between 18 and 25 watts, compared to the 10 to 14 watts from the two 6V6s in the Deluxe. Both the Mic and the Instrument channels used cathode-biased 6SJ7 tubes, which required a more elaborate circuit design than the 6SC7 twin triode, as on the Deluxe. The 6SJ7 *sharp-cutoff* pentode required two extra connections (screen and suppressor grids), but was capable of delivering more current (+50%) as well as running at a higher voltage (+20%) than the 6SC7. The sharp-cutoff feature of the 6SJ7 was the special winding of these tubes so that the amplification factor would go down the harder the input was hit. This was a compressor tube, designed to put out a high signal with minimal distortion from either a weak input or a hot one. The Professional and the Deluxe both used a 6N7 twin triode for their *paraphase* inverter, but the Professional was equipped with a 5U4 rectifier, which was capable of delivering more current (+80%) than the Deluxe's 5Y3. A heftier high-voltage supply included a choke (inductor) as well as larger filter caps. The components were still connected together where feasible, with ground connections soldered directly to the chassis. The control panel was the same as the Deluxe's and, in fact, had "Model 26" stenciled on, although the "26" was usually scratched out.

The control panel shows "26" scratched out; the Deluxe "Model 26" and Pro used the same panels.

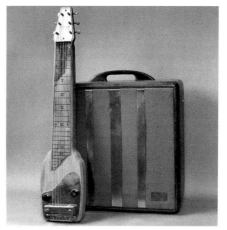
"Woodie" Professional (c. '47), '46–'47. "Professional" amp and steel guitar set.

Backside of amp shows Jensen 15" field-coil speaker.

Inside of chassis. *Note lack of circuit board.*

Bottom of chassis. *Note metal tubes and hefty transformers.*

TV-front Pro ('48), '47–'53. *King of the hill.*

Controls included Instrument and Mic Volumes and a backwards Tone (counterclockwise for full treble). The Professional was made in limited numbers, almost as if each one was built to order.

Shortly following the release of the Dual Professional (c. early 1947) the Pro, as it now was referred to, received a new look—the *TV-front* style, covered in vertical tweed. The new solid pine finger-joint construction was more of an important improvement for the large, heavy Pro than it would be for the smaller, lighter amps. The chrome chassis moved the controls from the back to the top, and an extra input was added. The tweed would be diagonal striped with a coat of lacquer by the release of the '49 catalog, a look that would remain basically unchanged until the release of the *wide-panel* Pro. The removable back panel covering the tubes and chassis was originally a solid rectangle in shape; around 1950 two vents were cut into the panels to let heat from the tubes, etc. escape.

Early changes to the circuit included replacing the 6SJ7s with two grid-resistor-biased 6SC7 twin triodes. This arrangement gave each input its own preamp, with all four the same; no more was there a hotter Mic input. The 6N7 paraphase inverter was replaced with a 6SC7, and some of the component values changed, but the power section was nearly the same. The choke was removed from the high-voltage supply (model 5C5). Later versions replaced the 6SC7s with 12AY7s in the preamp section, and a 12AX7 for the phase inverter.

The '53 foldout flyer showed the new wide-panel cabinet style (20"x22"x10"), but the amplifier really didn't change much. A Spare Speaker jack on the bottom of the chassis accommodated an additional speaker. A negative-feedback loop was added in an effort to reduce distortion, and the input tubes went back to cathode-biased. The addition to the control panel of a Standby switch allowed the high-voltage section of the amp to be shut off but left the heaters on, so that the amp would be ready to go with no warmup time (model 5D5).

Wide-panel Pro ('53), '53–'55. *Ad for Fender's top instruments, June '53. First stop for the Pro on its descent to mid line.*

In 1955 the Pro went to the new *narrow-panel* cabinet, keeping the same size. A Ground Switch was added to the control panel, as on all the big amps, but the previously top-of-the-line Pro still had only a single tone roll-off, whereas the Twin, Bandmaster, and 4x10 Bassman all had separate Bass, Treble, and the new Presence controls. A new design for the phase inverter circuit (*split load*) left half of a 12AX7 available as an extra preamp stage for all the inputs. The negative feedback was removed and, along with some other small changes, shows that the designers hadn't forgotten the Pro. But Fender really hadn't put the effort into the former King that it had for the other large amps. The circuitry in '55 (model 5E5) was still linked to the early models.

A new design for the Pro came around 1956 with the addition of Bass, Treble, and Presence controls. The negative feedback had to be returned to the circuit to install the Presence control, and this was not the only change. Instead of each of the four inputs having a separate section of 12AY7 preamplification, the two Instrument inputs shared half of one 12AY7 and the two Mic inputs shared the other half. This left both sides of the second 12AY7 to be used as additional preamp stages. The two sides were direct-coupled, with the second half a direct-coupled cathode follower feeding the tone controls—a big difference, as seen in late-fifties Bassmans (model 5E5-A). A fixed bias supply for the power tubes, which increased output, was made possible with an upgraded power transformer. The Pro would go through the rest of the fifties with no major changes, unlike the Twin and Bassman. The '58 catalog specified 5881s instead of the 6L6s and the change of the second preamp tube from a 12AY7 to a 12AX7.

With the new decade came big changes to the amps. The new brown Tolex, front-control-panel "Professional" series included the Pro (model 6G5),

Narrow-panel Pro ('55), '55–'60. *Still only one Tone control.... Came back in '56 with Bass, Treble, and Presence.*

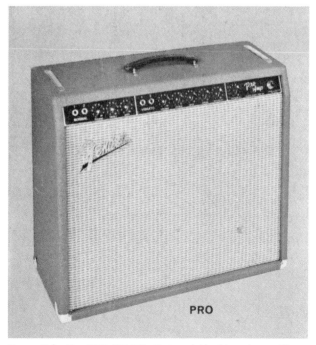

Brown Tolex Pro ('62), '60–'63. *First model had tweed-era grille cloth and "pinkish" brown Tolex ('60–'61), followed by maroon grille ('61–'62). Model pictured with wheat grille ('62–'63). Good looks...and great tremolo.*

which was nearly identical to the top-of-the-line Vibrasonic except for the speaker (and the price, almost $200 less). Front-panel controls were Bass, Treble, Volume for the Normal channel and Bass, Treble, Volume, Speed, Intensity for the Vibrato channel. Brown knobs and handle, chrome chassis straps, and a "flat" script logo completed the new look. The cabinet measured 20"x24"x10". Like the Vibrasonic, etc., the Pro used two 6L6 power tubes, five 7025s, and a solid-state rectifier. New to the Pro was a *long-tailed* phase inverter, first seen on the late-fifties Twins and Bassmans; these used both sides of one 7025. The Normal and Vibrato channels each had their own 7025 twin triode. Tone and Volume controls were placed between the two sides of these tubes, preventing any interaction from the other channel's controls. The last two 7025s made up the tremolo circuit, which was far more complex than previous designs. Operating only on the Vibrato channel's preamp signal, the two-stage oscillator varied the bias on two following stages, one for the high frequencies and the other for the lows. The sound of this tremolo was unlike any of the earlier models (see Vibrasonic).

By the end of '60, the controls would be rearranged to the more logical Volume, Treble, Bass order, and an extra 12AX7 was added to the revised tremolo circuit (model 6G5-A). Following the release of the maroon-grilled Showman, the Pro would receive the new cloth, replacing the tweed-era material. All the brown Tolex amps, by this time a different shade of brown, would change grille cloth for the second time in a little over a year (c. '62), this time to the wheat grille cloth. Otherwise the Pro would remain stable until mid '63.

A new Pro showed up in the *Down Beat* insert of September '63, featuring a blackface control panel, black-skirted knobs, and a black handle. The black Tolex cabinet was the same size as the brown Tolex models, and the flat logo was still being used, but the prototype solid gray grille cloth was unlike any earlier or later pattern. This Pro was also seen on the cover of the first-issue owner's manual, included with any amp Fender made from late '63 to '66.

Blackface Pro ('64), '63–'65. *The last of the 1x15 combos had a revised circuit with Bright switches replacing Presence control.*

The production blackface Pro had the standard silver sparkle grille cloth and the new raised logo, as seen in the '63–'64 full-line catalog. The amplifier itself had a number of changes from the brown Pros. The Presence control was removed from the negative-feedback loop, with Bright switches added in its place; Fender's new choice for the phase inverter, a 12AT7, replaced the 7025; a GZ34 rectifier replaced the silicon version; and the fixed bias supply for the power tubes now included a Bias Adjust pot. The all-tube tremolo section was replaced with one 12AX7 and a photoresistor, which dumped pulses of the Vibrato channel's signal to ground (model AA763). An AC outlet was added to the rear panel as a courtesy enticement to those wishing to add a tube Reverb Unit to their signal path. The blackface 1x15 Vibroverb, released almost the same time as the blackface Pro, should have been named the Pro Reverb. The Vibroverb was discontinued by the time of the CBS deal, and the 1x15 Pro would last only a few months longer, replaced with the 2x12 Pro Reverb. This ended the fading reign of the 1x15 combo.

The Pro Reverb (model AA165) was officially introduced in *Fender Facts* #9 (June '65), though earlier examples exist. However, there are *no* pre-CBS Pro Reverbs—Fender Electrics, yes; pre-CBS, no. The new cabinet (19½"x26¼"x10") housed the reverb pan at the bottom in a black vinyl bag and featured tilt-back legs and two medium-duty 12" speakers. The reverb section used a 12AT7 as the driver and one half of a 7025 as the return. The Normal input ran through two preamp stages of 7025, but the Vibrato channel ran through three, not to mention the three extra stages from the reverb circuit. Much of the circuitry was unchanged from the 1x15 model. One only has to play through a blackface Pro Reverb to realize it would be a few years after CBS' takeover before a noticeable difference in quality of sound occurred. If you happen to use a Fender guitar or other brand with single-coil pickups, the author recommends these amps as the best all-around amp ever made—by anyone. (Use a humbucker? See Vibrolux.)

Blackface Pro Reverb ('65), '65–'67. *A Pro without a 15" speaker? This amp became a classic, nicknamed "the Baby Twin."*

Silverface Pro Reverb ('72), '68–'81. *Early models fitted with aluminum trim ('68–'69). Model pictured has no Master Volume. Built-in casters at no extra charge. Last models returned to blackface ('81–'82).*

Nineteen sixty-eight saw a dressed-up version of this amp, identical except for its new silver control panel and blue grille with aluminum trim. Then came the changes, in particular to the bias section (model AB668). Capacitors off the grids of the power tubes and resistors off the cathodes were not sonically appropriate. The cathode was soon put back to ground (model AA1069), but the capacitors remained, reportedly to squelch high-frequency feedback caused by moving all the wires that previously ran under the circuit board to the top. On paper it shouldn't have been necessary, but in real life the placement and even the horizontal or vertical alignment of components can change the way an amplifier operates. A 5U4 replaced the GZ34 at this time, and the aluminum trim was dropped, as seen in the 1970 catalog. A three-position ground switch was the only new feature. The cabinet was enlarged to 10½" deep, and by '72 it was narrowed to 26", so that it could share cabinets with the Twin. Casters became a standard feature around this time.

A Master Volume circuit was added in the mid seventies, which also included a Boost, or distortion, activated by the pull-knob potentiometer. The logo by this time lacked the underlining "tail," and a Hum Balance control was added to the rear panel. The power was reportedly increased to 70 watts in the late seventies, and a midrange—or Middle—control was added to the Vibrato channel. The combination of blackface and silver grille was again offered in 1980 and soon became standard appointments on the last of the Pro Reverbs, as seen in the 1982 catalog. The accompanying caption stated, "A rugged, reliable work horse that's been a favorite for a long time—and will be for a long time to come." The model was discontinued later that year. A solid-state amp bearing the name Pro 185 was released in '88 (see page 153).

SUPER
1947-1982
1988-Present

Dual Professional
Super
Super Reverb
Super 60
Super 112
Super 210

One of Fender's shortest-lived models would also be one of its most influential, both for future Fender amplifiers and for the rest of the industry. If nothing else, the Dual Professional introduced tweed to the line at a time when the three original amps were still uncovered hardwood boxes. The chromed chassis with top-mounted control panel would soon become standard construction for all but the smallest amp (which would eventually join the others). Another feature, surrounded by an aura of mystique, is the use of twin speakers, a first for Fender and generally considered an industry first as well.

Specifics of the Dual Professional include twin 10" Jensen PM10-C speakers and twin output transformers, mounted directly to the speakers. Fender may have used one for each speaker because of power-handling capabilities or for impedance considerations. The amplifier, similar to that of the mighty Pro, used two 6SJ7s for the preamp, a 6N7 for the phase inverter, two 6L6s for the power, and a 5U4 for the rectifier. It was the first amp to have a tube chart mounted inside the cabinet. A Mic input with a level control and two Instrument inputs with a separate level control were equalized by a common Tone roll-off circuit. The insides of the amp were exposed when the back panel was removed, making service of everything but the transformers possible without removing the chassis. Components were mounted to a circuit board—a Fender first! An On/Off switch, a fuse holder, and a red pilot light were mounted on the face of the chassis, which was chrome plated and labeled in white. A "Lo Gain" jack was a short-lived feature, unique to this model, that allowed a high-output device to be "inserted" into the signal path after the preamp tube and coupling cap, just before the Instrument Volume and the master Tone. The modern-day equivalent would be "Power Amp In." A lightning-bolt logo was positioned between the pilot light and the tone control.

Dual Professional, c. '47. *Fender's first tweed. Its successor, the Super ('47–'53), debuted in a similar V-front cabinet.*

Fender's first top-mounted control panel. Note Lo Gain input.

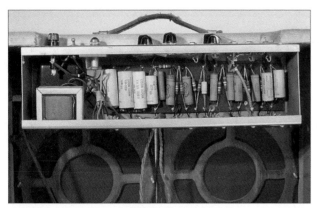
Back panel removed to expose Fender's first circuit board. Note intricate cutout of baffle board.

Twin speakers—another first for Fender, and possibly for the industry. Note dual output transformers and baffle boards. Speakers date from the thirty-second week of '46.

Wide-panel Super ('54), '53–'55. *New cabinet with single baffle board, styled after Twin.*

The cabinet was made of solid pine, using finger-joint construction, much stronger than the "woodies," but having unsightly corners—another reason for the tweed covering. The first tweed material Fender used had a pattern that ran straight up and down and was a uniform color, very light, with a thicker weft than warp. The grille cloth was a dark brown mohair that contrasted nicely with the almost white tweed. Setting the whole look off was a metal strip running vertically down the center of the grille. Besides looking very stylish, it held the two baffle boards together. A small metal name tag with "Fender/Dual Professional/Fullerton California" was tacked to the front of the cabinet, and a leather handle topped off the package.

The Dual Professional became the Super by the fall of '47 as the tweed became a two-tone brown, still running straight up and down. The nameplate was changed to a generic badge reading "Fender/Fullerton California," and the lightning-bolt logo was replaced with "Super-Amp" in white script. Blue-frame Jensen Concert Speakers replaced the original models, and the amp changed to a single output transformer. The unique look of the 2x10 amp would go unchanged until '53 except for the change to diagonal tweed, as seen in the 1949 catalog. The size of the cabinet, however, was increased slightly at least twice. Like the Pro, early changes to the Super's circuit included replacing the 6SJ7s with two grid-resistor-biased 6SC7 twin triodes, and the 6N7 *paraphase* inverter with a third 6SC7. A fourth input was added, and at this point all the inputs had identical gain, although the Instrument inputs were brighter (at least at low volumes). An extra filter cap would be substituted for the choke, and the last examples of this circuit (model 5C4) used 12AY7s for the preamps and a 12AX7 for the phase inverter.

The angled front must not have proved worth the effort and expense, as a cabinet with a flat front and a single baffle board (18"x22"x10") replaced the original style around the time the Twin Amp was released. The new *wide-panel* Super was soon equipped with a Standby switch and Spare Speaker jack, and the preamp tubes changed from grid resistor to cathode biased. A *self-balancing paraphase* inverter circuit was another improvement, albeit short-lived (model 5D4).

The Super for 1955 featured the new *narrow-panel* tweed look and a nameplate with a script logo. Grille cloth was really grille cloth, acoustically transparent, with light and dark brown threads. The speakers were mounted off center, in the lower left and upper right corners of the new cabinet (18½"x22"x10½"). Up to this point, the Super and the Pro were basically the same except for their speakers, and the narrow-panel models would continue the tradition. Both models went from three controls to five, including two Volume controls and an expanded tone circuit with separate Bass, Treble, and Presence knobs. Instead of *two* 12AY7 twin triodes for the inputs (one section for each of the four inputs), a new circuit used half of a single 12AY7 for each channel. The two Bright inputs (High- and Low-gain) were summed together using 68kΩ resistors, and the two Normal inputs followed suit. This became the standard input section for Fender, even after the preference for preamp tubes changed to 12AX7s in the sixties. Although both the Super and the Pro used the same preamp tubes as earlier models, the way they were used changed dramatically. The second 12AY7 now was common to both channels, and the two sides were direct-coupled, with the second stage a cathode follower preceding the tone controls. Half of the old 12AX7 phase inverter tube was used as a third gain stage, with the other half handling the *phase splitter* function. A fixed bias supply for the power tubes replaced the less efficient cathode-bias arrangement of previous amps, and a choke was added to the filter section. A ground switch was added to the control panel. The power section on the Super was down-sized to two 6V6 tubes for a short time (model 5E4-A), although the plate voltage for the power tubes exceeded the rating of 6V6s by 25% and was identical to the Pro's, which was operating 6L6s. A return to the 6L6s in '56 (model 5F4) was accomplished simply by changing the value of the resistor in the fixed bias section. This upgrade was about all that was done to the amp during the rest of the fifties.

Narrow-panel Super ('55), '55–'60. *New model featured Bass, Treble, and Presence controls. Note speaker alignment.*

Brown Tolex Super ('60), '60–'63. *Brand-new amp with front control panel, tremolo, more power, etc. Second and third versions with maroon and wheat grilles.*

Blackface Super Reverb ('65), '63–'67. *A favorite of many since its inception..."brings back by popular demand the tonal characteristics of the old-style Bassman Amp plus built-in Reverberation and Vibrato."*

The new Professional line, outfitted with brown Tolex, front control panels (with brown knobs), and tremolo, included the Super as its lowest-priced model. All the amps had two 6L6 power tubes, five 7025 preamp tubes, and silicon rectifiers—except for the Super, which was fitted with a GZ34 tube rectifier (model 6G4). Two independent channels (Normal and Vibrato) were each fitted with (L to R) two Input jacks plus Bass, Treble, and Volume controls. The Vibrato channel also had Speed and Intensity controls, and a Presence control operated on both channels. The 1960 model still had the grille cloth from the late-model tweeds, which was replaced in '61 with the maroon cloth first seen on the Showman. The 7025 operating the tremolo circuit was replaced with two 12AX7s in '61 (model 6G4-A), and the Tolex was now a slightly different shade and texture. It's almost necessary to see the two side by side to notice. In '62 all the Professional-series combos joined the small combos in wheat grille cloth, which also replaced the maroon grilles on the piggybacks by the end of the year.

Early in '63 Fender released the reverb-equipped Vibroverb, a catchy moniker for what was essentially a Super with reverb, and the name game was on. Later that year the 2x10 Super was replaced with the blackface 4x10 Super Reverb, which was essentially a Concert with reverb. The brown Vibroverb changed to a blackface 1x15 Vibroverb, essentially a Pro with reverb. By early '65 the Concert and Vibroverb were gone, and the 1x15 Pro was replaced by the two-6L6, 2x12 Pro Reverb. The two-6L6, 1x12 Vibrolux was retired, replaced by the new two-6L6, 2x10 Vibrolux Reverb, which was essentially a brown Vibroverb, which was essentially a brown Super. The 4x10 Super Reverb is a classic amp, but has closer ties to the brown Concerts of the early sixties and the tweed Bassmans of the late fifties than to the earlier Supers. This is even mentioned in mid-sixties descriptions of the Super Reverb.

The Super Reverb amps have been favorites for years with blues and rock players in particular, who favor the gutsier two-6L6 sound to the crystal-clear sound of the Twin Reverb. Having four 10" speakers gives the amp a nice tight bass response, and the blackface circuitry into 10" speakers puts out enough treble "to take your head off," as they say. Internally, the blackface Super Reverb (model AA763) was considerably different than either the Super or the Concert that preceded it. Gone from the "brown" days was the tube-oscillator tremolo circuit, replaced with a photoresistor network. The Presence control was omitted, but the amp gained a Bright switch on each channel in its place and a "Middle" control on the Vibrato channel. Most important, though, was the addition of reverb. Both sides of a 12AT7 were used to drive the reverb pan, which returned into a 12AX7. The Normal and Vibrato channels each used a 7025 as a preamp, and all this was fed into a 12AT7 phase inverter, followed by a pair of 6L6s. A GZ34 was used as the rectifier, and a 12AX7 was used in the tremolo circuit.

Hitting production just behind the new Twin Reverb and Deluxe Reverb amps, the Super Reverb missed joining them on the cover of the new Operators Guide (owner's manual), and also in the *Down Beat* insert of September '63. It was finally introduced in the '63-'64 full-line catalog.

Silverface Super Reverb ('76), '68–'81. *First version fitted with aluminum trim ('68–'69). Later models equipped with Master Volume and optional JBL speakers. Note alignment of speakers, instituted c. '69. Last models returned to blackface ('81–'82).*

The cabinet was enlarged from 24"x24"x10½" to 25½"x25½"x9¾" around the time the silver panel with aluminum trim and blue sparkle grille cloth were added, c. 1968. For a short time a smaller raised logo, minus the tail, was added in place of the regular style, but the standard logo soon returned. The small tail-less logo differs from the later full-sized version, lacking the "®" symbol. Shortly after the introduction of the silverface amps, the speakers were no longer mounted in the four corners. Instead, the top two were "slid" left a bit and the bottom two "slid" right. The rear control panel received a Hum Balance control, and the bias circuit underwent major changes in an attempt to further "clean up" the sound (model AB568). Most of these changes can be undone to return the amps to a blackface circuit. Fender actually made a few of these reversions themselves the following year, upon realizing that they had gone astray (model AA1069). The 1970 models lacked the aluminum trim and went to a slightly smaller cabinet (24½"x25"x10"). A three-position ground switch became standard around this time, and casters were added at no extra charge by 1972. The Super Reverb managed to

Super 60 ('89), '88–'92. *Fender's only two-6L6 amp at the time.*

postpone the addition of a Master Volume control with pull-knob Boost until the mid seventies. The logo lost its tail about then, and power was boosted to 70 watts.

Little was done after this, the most obvious thing being the return of blackface cosmetics, first as an option in 1980 and then as standard issue around a year later. A Line Out jack was added to the back panel, and a Middle control was added to the Normal channel. The Super Reverb, long one of Fender's most popular combos, was phased out in 1982. The 4x10 Concert, part of the new-for-'82 "II" series, was for all practical purposes an extension of the Super Reverb.

Following the changeover from CBS in '85, the new company revived the Super name in '88 for the 60-watt Super 60, with red knobs, one 12" speaker, and, of course, no tremolo. The controls were similar to those of "The Twin."

Super 112 and 210 ('91), '90–'92. *Designed to replace the Super 60, which, due to popular demand, ended up staying in the line beside them. Phased out with the introduction of the Professional tube series.*

Two new-style amps, the Super 112 and 210, not to mention a rack-mount version, joined the Super 60 in '90 and were replaced in '93 with a retro-looking all-tube 4x10 "Super," a counterpart to the 1x12 Concert in the Pro Tube series. The Super and the Concert are identical, with the exception of their speaker configurations and the size of their cabinets. (See Concert for details of the electronics.) Even the output transformers are the same, with the four 10s wired in *series-parallel*, yielding an 8Ω load instead of the 2Ω usually associated with 4x10 Super Reverbs wired in parallel. Possibly the best all-around amp made today if you require channel switching and don't need tremolo.

Super ('93), '93–. *The wonderful combination of two 6L6s and four 10s.*

CHAMP
1948–1994

Champion "800," "600"
Student
Champ
Vibro Champ
Champ II
Super Champ
Champ 12
Champ 25SE, 25

Champion "800" ('48), '48–'49. *The first model in Fender's long-running Champ line.*

Logo on control panel.

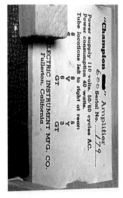

Early Champion "600s" came with "800" tube charts with the old model number inked out.

Two-tone Champion "600" ('52), '49–'53. *Champion promotions were geared to teachers and students.*

With the upgrading of the Princeton to include Volume and Tone controls c. 1948, room was made at the bottom of the line for a new series of amplifiers. The Champ line has offered rugged and dependable packages at reasonable prices ever since, aiming at the entry-level student market, but also being a favorite for recording artists due to their minimal circuitry.

The first model was the Champion "800," featuring approximately 4 watts of single-ended Class A power, an 8" speaker, two inputs, and a Volume control, all housed in a solid pine box. This was covered in "green" tweed linen and fitted with a leather handle and a nameplate bearing the Fender logo. The control panel was "finished in a very attractive, gray-green hammerloid finish with white marking." The green color scheme was unique to the short-lived "800," but the logo was common to all the Fender amps, giving this midget an element of credibility. The three tubes—6SJ7 (preamp), 6V6 (power), and 5Y3 (rectifier)—are to this day all considered excellent sounding, though not necessarily the most powerful. These amps were offered at least from the summer of '48 through the spring of '49.

By June of '49, the "800" was replaced with the down-sized Champion "600," having an identical circuit and even using the tube chart of the "800" on the earliest models. The "600" was covered in two-tone (brown and white) textured vinyl and equipped with a 6" speaker, a complement to the Princeton's 8". This covering was again unique to the Champion, although the white vinyl has shown up on a few amps of the early fifties, in particular a *TV-front* Bassman (see page 182) and a TV-front Pro. (The brown was used on Tele cases for a short time.) This two-tone style would run until 1953. The rear-facing panel would last until the change to the *narrow-panel* cabinets, c. 1955. These control panels were simply a bent piece of sheet steel with the components mounted directly to the metal, *sans* circuit board. On all but the earliest models the panel/chassis was copper plated, with the outside face (control panel) painted brown with white lettering. This sheet of metal holding the tubes, transformers, controls, etc. slid into channels in each side of the cabinet and was secured to the back with two roundhead screws.

Chassis/control panels slid into grooves in sides of cabinets. "800" (left) and "600." Note lack of circuit boards.

With the introduction of the *wide-panel* box, the Student Amp, as it was often referred to, joined the rest of the amps cosmetically. Now covered in tweed, the earliest version of this new style still had the "600" logo on the control plate; later ones had "Champ-Amp." These had model numbers 5C1 followed by 5D1, although there was no direct correlation between the names and the model numbers.

Wide-panel "600" ('54), '53–'55. *Same amp as two-tone, but with new wide-panel tweed styling. Note ported back panel. Last models say "Champ-Amp."*

Around 1955, the previously angled rear control panel/chassis was replaced with a standard Fender chassis and circuit board, and the controls were moved up to the top. A 12AX7 preamp tube was standard by this point. The wide-panel box was replaced with the narrow-panel style (11"x12"x7¼"), and the speaker opening was covered in the new "genuine" grille cloth. In keeping with the larger amps, a choke was added to the high-voltage filter section, and the use of negative feedback was employed (model 5E1). This circuit design would remain virtually unchanged for almost 10 years, except for the removal of the choke the following year (model 5F1).

Narrow-panel Champ ('55), '55–'64. *New style with top-mounted control panel, regular back panels, and new grille cloth. Adopted larger box c. '56 and 8" speaker c. '57.*

A new, larger box (12½"x13½"x8") was added c. 1956, and the speaker was upgraded to a single 8" a short time later, ending the development of the tweed Champ. The Champ was the swan song of the production-model tweeds, lasting until 1964, when the new black Tolex, silver grille cloth, and black plastic handle were added to the last of the boxes with chrome chassis and top control panels. This short-lived hybrid of old and new, with its lone black pointer knob, was Fender's last cosmetic tie to the 1948 series.

By fall 1964 the Champ was totally revamped (model AA764), gaining separate Bass and Treble controls and a front-mounted blackface control panel. The tubes were unchanged, and the layout of this version would become a standard, being left alone until 1982 and selling by the

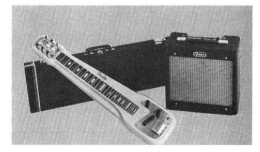

Transitional Champ, '64. *New black Tolex and silver sparkle grille on tweed-era cabinet, still with chrome control panel and pointer knob.*

FENDER AMPS: THE FIRST FIFTY YEARS

Vibro Champ ('70), '64–'82. In fall '64 the Champ changed to a front control panel, with separate Bass and Treble controls. A similar model, the Vibro Champ, featured tremolo. The first of these amps were blackface ('64–'67), followed by silverface with "blue" grille and aluminum trim ('68–'69). Model pictured ran basically unchanged until 1982.

Champ II ('83), '82–'83. This short-lived powerhouse put out a whopping 18 watts (as compared to four and a half on previous models) into a 10" speaker.

Super Champ Deluxe, '82–'85. The Super Champ was similar to the Champ II, with the addition of channel switching and reverb. The limited-edition Deluxe version shown here featured an E-V speaker; solid oak cabinet; brown grille cloth, knobs, and control panel; leather handle; and came with a padded cover.

thousands. A souped-up version, the Vibro Champ (also model AA764) was added as well, featuring built-in tremolo with separate controls for the speed and intensity. An extra 12AX7 was added to the circuitry for the tremolo. The Champ would follow the rest of the amp line in changing to the silverface panel with "blue" grille cloth and aluminum trim by 1968, losing the trim by 1970 and adding a new tailless logo in the mid seventies. In 1982 the Champ and Vibro Champ were again outfitted with blackface control panels, albeit with a different style of silk-screened lettering. These were the last of the, by this time, 6-watt wonders.

Nineteen eighty-three saw the introduction of the single-input Champ II, featuring a 10" speaker and completely new circuitry, including Master Volume and a Bright switch. Two 6V6 power tubes generated a whopping 18 watts, making this little package surprisingly loud for a Champ. A fancier version, the Super Champ, included reverb and a hot-rod lead channel. "A new legend in the making!" was the company's prediction, and it has held true, as this short-lived variant almost immediately became sought after by those looking for a very small, loud, full-featured amp with good tone. A special-edition Super Champ Deluxe came equipped with a 10" Electro-Voice speaker, a lacquered solid oak cabinet, and special grille cloth, nameplate, and cover. These genuinely were special editions. The Champ II was quickly dropped, but the Super Champ survived until the sale of the company by CBS. Existing inventory was sold through 1986.

A wall of Champs, c. '48–'85. (L to R) Bottom row: '48 "800," '49 two-tone "600," '53 wide-panel "600," '55 small-box narrow-panel Champ, '56 big-box narrow-panel Champ. Second row: '65 blackface Vibro Champ, '65 blackface Champ. Third row: '83 Super Champ Deluxe, '83 Super Champ. Top: '64 Champ with Tolex on a tweed-era cabinet.

The first small amp from the new Fender Musical Instruments Corp. was the "Red-Knob" Champ 12. A 12" speaker and 12 watts of power qualified the name, and tape inputs, a line output, and a headphone jack made the Champ 12 convenient for learning songs off tapes, home recording, and silent practicing. Five custom coverings were available, including red, white, gray, and snakeskin!

The hybrid Champ 25SE replaced the all-tube 12 and featured two 5881 power tubes, two channels with independent tone controls, and a totally restyled package featuring black grille cloth, plastic corners, old-style logo with tail, and black knobs. Another version, without the tape inputs, line out, and master volume control, was called the Champ 25. These were really too big and complex to be considered Champs. Fender discontinued the model in 1994 after a short run, leaving the company without a tube Champ for the first time since 1948. Blasphemy!

Champ 12 ('91), '87–'92. *First tube amp for the new owners.*

Champ 25 and 25SE ('94), '92–'94. *A cross between a tube Champ and a mid-sized solid-state amp, the short-lived 25s used the hybrid approach. The last Champ with a tube.*

BASSMAN
1952–1983
1990–Present

> *Bassman*
> *Super Bassman*
> *Bassman 50*
> *Bassman 10*
> *Bassman 100*
> *Bassman 70*
> *Bassman 135*
> *Bassman 20*

Which came first, the Bassman or the bass? Probably the bass, with prototype Precisions showing neck dates as early as October 1951. Fender's foldout flyer from '52 officially introduced the P-Bass, paired up with the *TV-front* "Amplifier." From the front this looks like a standard 1x15 Pro of the time, but the back was unlike any other Fender amp, before or since. The most obvious difference was the sealed back, with a cutout at the bottom for a metal panel (housing the power cable, fuse, on/off switch, and pilot light) and two round "ports" halfway up the cabinet. (In a very dark room, the glow of the tubes coming through these openings gives an eerie jack-o'-lantern effect!) The only specs given were "special 15" Jensen bass speaker, unique speaker housing for bass emphasis." Removal of the back panel on "Bassman Amp" serial no. 0155 exposes the chassis, made of copper-plated steel, mounted to the bottom of the cabinet, and a control panel, also of copper-plated steel, mounted to the top. Two inputs, a Volume control, and a Tone roll-off pot are the only components on this piece of metal, finished in brown with white letters on the exposed side. A five-wire "snake" runs

TV-front Bassman ('52), '52–'53. *The first Bassman. Note the ported back.*

from the panel, around the speaker, to the back of the chassis, where it is secured with an octal tube socket. A ¼" output jack for the speaker is the only other connector on the back. The 6SC7 preamp tube is split, with one half going to one input jack and the other half to the second input jack. A 6SL7 *paraphase* inverter feeds two 6L6 power tubes, and a 5U4 rectifier tube turns on the DC. Unfortunately, the Jensen Concert Series 15" in this amp is dated 1954, implying a blown speaker at one time. The pots appear to be original, with dates of the seventeenth and forty-first weeks of '51. This implies the possibility of a very, very late '51 manufacture, but '52 seems more reasonable and is generally accepted as the first year for the Bassman. The first P-Basses were reportedly shipped with TV-front Pros.

Ads for the new bass-and-amp pair began running in the trade magazines, as well as *Metronome*, *Down Beat*, etc., in '52. The foldout flyer from '53 featured the new *wide-panel* styling, as did ads starting that year. Although the photo of the backside of the amp shows an open-back cabinet, close examination shows three controls, while the text and schematics specify only two. The photo is probably the same shot used for the three-knob Pro and four-knob Bandmaster, which were of the same dimensions (20"x22"x10"). The '54 full-line catalog shows a repeat of this shot. The wide-panel Bassmans (Bassmen?) had the same basic construction as the TV-front models. Serial no. 0381 is an example of this style. Again, the speaker has been changed, but a date code on the tone pot shows the thirtieth week of '52, implying an issue date of very, very late '52 into '53.

Just as the Dual Professional and the Twin had necessitated wide-panel cabinets, the Bassmans that followed required a new-style cabinet. The single speaker was replaced by not a pair of speakers side by side, but *two* pairs of speakers side by side, stacked one pair above the other. The *narrow-panel* 4x10 Bassman was the result, and it supposedly came about because of the earlier models' inclination to blow 15s! The four 10s did the job well; many of these

Wide-panel Bassman ('53), '53–'54. *Same as TV-front, except for new styling of front panel.*

'52–'54 Bassman with back panel removed. *Note chassis on bottom and "umbilical cord" to controls.*

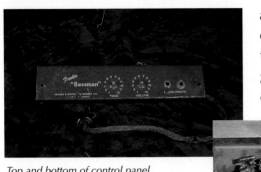

Top and bottom of control panel.

amps are still found with original blue Jensens in fine working condition. And these amps usually have been played *a lot*, as this model has been a favorite for years (more often with guitarists than bassists). From Fran Beecher with Bill Haley, and Bo Diddley with his two extra speakers (pointing backwards!), to Bruce Springsteen and Jimmie Vaughan, from recording studios that keep one as their "house" amp to the weekend warrior who escapes the Monday-through-Friday grind answering to no one but his guitar and amp, the 4x10 Bassman is to this day a sanctuary of tone for many. Those that desire one give up whatever it takes to obtain their holy grail.

But not all 4x10 Bassmans are alike. Appearing on the February '55 price list, the first models (5E6, reportedly also 5D6) had only two inputs: one Normal and one Bright. The company's designers obviously intended this amp to sound good for guitar as well as bass, conveying in the '55 full-line catalog that "While its characteristics have been designed to accommodate string bass, at the same time it makes an excellent amplifier for use with other musical instruments." No truer words could be written.

Inside of chassis. Note octal connector and ¼" speaker jack.

These amps used two rectifier tubes wired in parallel to decrease *sag* (a drop in the high voltage) when the amp was pushed. Solid-state rectifiers accomplish the same thing, and today this is a point of contention between players preferring one sound to the other. For those preferring a compressed sound, the removal of one of the rectifier tubes is a step in that direction, giving more sag. Fender claimed these amps were capable of producing 50 watts of "high fidelity audio power" from two 6L6 power tubes with two 12AY7s and a 12AX7. Separate Bass, Treble, and Presence controls were a major improvement over earlier single Tone roll-off controls, and Standby and Ground switches were also new.

The cabinets (23"x22½"x10½") had a new grille cloth on both the front baffle board and the cutouts on the bottom back panel. To fit four speakers in such a small area, the bell-caps (magnet covers) had to be removed from the top two Jensen

P10-Rs. A new nameplate with "Fender Bassman" in script replaced the earlier block-letter version, and a leather handle topped off this basically all-new amp. The price was now the same as Fender's previously top-of-the-line Twin Amp.

In '57 a new, improved Bassman was unleashed, having four inputs (high- and low-gain Bright and high- and low-gain Normal) and yet another tone control: Middle—"another Fender First." The six-knob Bassmans (and Twins) were vastly more versatile than the amps with a single tone roll-off of less than five years before. A number of other circuit changes at this time (model 5F6) included a new single rectifier—the mercury-vapor 83, with virtually no sag—more gain from swapping the second 12AY7 for a 12AX7, a new phase inverter circuit, 5881s (which at the time were actually a better-quality 6L6) in the power section, and higher plate voltages.

In '58, a slightly different version (5F6-A) replaced the discontinued 83 rectifier with a GZ34 (a.k.a. 5AR4). In the mid fifties Fender had added a 1500Ω "swamper resistor" between the grids of the power tubes and the bias circuit to help stabilize the amp; "to damp out any tendency for ultrasonic oscillation" is the technical explanation. This addition was made on all the push-pull amps during the fifties and has been used somewhat regularly by Fender ever since. The 5F6-A Bassmans had these removed, another possible reason why this last version of the tweed 4x10 Fender is considered by many to be "the one." Perhaps the highest compliment of all is that it was this model that Jim Marshall and Ken Bran chose as the pattern for the first Marshall amp.

Narrow-panel Bassman ('60), c. '55–'60. *The 4x10 Bassman ended up as one of the greatest guitar amps of all time. Pictured is final and most revered version, reissued in '90.*

The leather handle, which had a tendency to break on the heavy amps, was replaced in '59 by a new brown molded plastic "Fender" handle, which also had a tendency to break on everything it was on. The reissue version of this handle is of a more pliable rubber and works quite well, although it is a bit snug on some models.

Despite having been pushed previously as an amp for every

need, the 4x10 Bassman was not part of Fender's 1960 "Professional" series of amps that featured brown Tolex, front control panels, tremolo, and brown knobs. The new 4x10 Concert was obviously trying to woo the Bassman-playing guitarists. Rumors of brown Tolex 4x10 Bassmans exist, and although never mentioned in official company promotions, the idea should not be discounted. (The author examined one a few years back, and at that time was convinced it was original. Since then, stories from dependable sources of a brown '60 Bandmaster and a brown '60 Twin, both with tweed-era top-mounted chrome chassis, have supported this belief.) The production-model 1960 Bassman, however, was the last of the big tweed amps.

The sixties saw the Bassman in a continuous state of change. Between 1960 and 1970, Fender issued 10 up-to-date, full-color, full-line catalogs, with no two Bassmans alike! Granted, some of the changes were less significant than others, but the list also needs to include two that definitely were production models but that were issued between catalogs.

Following the December '60 release of the new Showman, the February '61 price list showed the Bassman as a "Piggyback Unit," "Textured Vinyl Covered," with a "12" Heavy Duty Speaker." The Bassmans and other models with white Tolex and maroon grille cloth are *objets d'art* and were chosen as the basis for the new Custom Amp Shop models. The contrast between the dark grille, brown control panel and brown handle (as seen on late-fifties Bassmans), and white Tolex and white plastic cylinder knobs has made this series a favorite for years. (Did I mention that these make great guitar amps?)

The '61 Bassman (model 6G6) was essentially a brand-new design, with Bass and Normal channels, each with Volume, Treble, and Bass controls, and a master Presence control. The GZ34 and 5881s were holdovers from the tweed era, as were the basically unchanged phase inverter, bias, and power sections. The four 7025s were a new combination, with the Bass channel having two of the twin-triode 7025s as preamp stages—essentially a total of four. The first 7025 had its two sides direct-coupled, with the second half set up as a cathode follower. While all the narrow-panel tweed amps used this arrangement in front of the tone circuitry, the white-knobbed Bassmans were the only amps of the Tolex

Piggyback Bassman, white Tolex with maroon grille ('61), '61–'62. *Grille cloth changed to wheat late '62 to late '63. A long-time favorite guitar amp for many, one of the all-time best!*

era to continue the practice. Besides the standard negative-feedback loop that included the Presence control, a second feedback loop was added around the fourth preamp stage. This last stage also had an unbypassed cathode resistor, causing the tube to have slightly less gain, but a more even frequency response and less distortion. The Bass channel on these amps is unique in its circuitry and a long-time favorite of—again—guitarists who could hear the difference. The Normal channel was normal Fender, having only two gain stages, from one 7025.

Dimensions of the head were 8"x22¼"x9". The cabinet measured 21"x30"x11½" and was fitted with "tilt-back" legs and two T-nuts mounted in the top, which allowed the head to be strapped down to the cabinet. The single 12" speaker was mounted in a special ported cabinet that greatly improved bass response (see page 110). It's possible Fender again had trouble with using a single speaker for reproducing the bass, because the ported 1x12 cabinet was replaced with a sealed 2x12 version (21"x32"x11½") to spread out the power and damp the speaker excursion. This model is pictured in the '62–'63 full-line catalog. (Only the Showman, equipped with a single J.B. Lansing speaker, would keep the ported cabinet.) The tube rectifier was replaced with a silicon rectifier during the era of the maroon-grille 2x12 (model 6G6-A), and these amps usually have a cap over the hole that previously held the rectifier tube socket. The rest of the circuit was virtually unchanged. The negative feedback in the preamp section was removed, as the combination of this and the unbypassed cathode resistor was a bit overboard. But the feedback in the power tube section was increased a bit.

In late '62 the maroon grille cloth was replaced by the wheat-colored variety used on the combo amps. A black reinforced handle replaced the brown "Fender" handle in early '63. This version lasted into the middle of '63, although neither wheat-cloth/white-knobbed amp was catalogued.

The '63 full-line catalog showed a couple of piggybacks that were never production models: an early-'63 Tremolux and an early-'63 Bassman, both fitted with the new skirted black knobs. These were installed right over the numbers on the brown panels. The amps pictured are lacking the new Bright switch as seen on the Bandmaster, Showman, and new

Imagine this amp with white knobs and you have the actual production model for mid '63. The black knobs were added for the catalog photo.

FENDER AMPS: THE FIRST FIFTY YEARS

Bassman, black panel with white knobs, mid '63 to late '64. *Most of these had white Tolex with gold sparkle grille (see page 197). During the last few months the amp went to black Tolex with silver grille, as seen here. Still with old circuit.*

Blackface Bassman with black knobs ('65), '64–'67. *A revised circuit with Bright and Deep switches instead of Presence. Changed to BIG cabinet in '67. Silverface Bassman became Bassman 50 c. '72, followed by Bassman 70 c. '77–'83.*

blackface combos. The Tremolux was soon updated to a standard numberless blackface control panel, with black, numbered knobs and the Bright switch, but the Bassman kept the old white knobs and Presence circuit alive for another year. The old numbered control panels were now black instead of brown, the flat logo was replaced with the raised logo, and a new gold sparkle grille replaced the last of the wheat cloth. This model also didn't appear in any catalogs, because the white Tolex (by this time smooth-textured) and gold sparkle grille were replaced in the middle of '64 by black Tolex and silver sparkle grille, as seen on the blackface models. These black Tolex Bassmans with white knobs (a very sharp look, also seen on the Princetons of late '63 to mid '64) appeared just in time to make the '64 full-line catalog. They were almost immediately replaced by the new black-knobbed blackface style. The circuitry on this model (AA864) was significantly different than that of the preceding models. A return of the 1500Ω swamper resistors, the removal of the Presence control, and the addition of Bright and Deep switches were a few of the obvious changes. The Bass channel (which we like so much for guitar) had one of its preamp stages removed, leaving one half of a 12AX7 twin triode not connected—just sitting in the socket. This is a highly respected blackface Bassman, designed and built before CBS was involved. Fender was simply looking for what they considered a better sound, which was clean and rich. The fact that certain amps sound the way they do when pushed to their limits was not the primary concern of their designers. The early amps were primitive; by the mid fifties they had reached the necessary levels of power and tonal variety; and in the early sixties the workings and response characteristics were being refined. Some people prefer the tweeds, some the white amps, and some the blackface models. They're all justifiable preferences.

The last of the pre-CBS Bassmans would have the letters "NL" stamped on the tube chart; all the "O" models are CBS, even if they still say "Fender Electric Instrument Company" on the control panel, as seen in the 1965 catalog. The last of these have a slightly different bias circuit (model

AA165), but not the notorious one, which was yet to come. The '66 catalog showed the same amp—and it is the same amp, except for the new "Fender Musical Instruments" designation. This change actually occurred in the summer of '65, and although this is a petty detail, there are those who believe in Mojotonal superiority and that the Fender Musical Instruments models have been hexed. Not true!

A giant 2x12 bottom was unveiled in the 1968 catalog, which should say '67–'68, as it was prepared in the summer of '67. The 40"x29½"x11½" cabinet, over 75% bigger than the earlier Bassman bottom, still housed two 12" speakers, although one would think it had four 12s—or six 10s—which is probably the impression it was designed to give! In the upright position it was taller than a Marshall half stack, although at that time Marshall posed no real threat to the mighty Fender line. Three handles were mounted on one side so it could be laid down, and casters were on the other. The top still had the T-nuts to secure the head to the cabinet, but the tilt-back legs were gone.

The '69 catalog, really the '68–'69, finally showed the new silverface models with "blue" grille cloth and aluminum trim. This theme was introduced on the solid-state line in the summer of '66, but would not be used on the tube amps till more than a year later. The circuitry on the silverface Bassmans of early '68 was the same as on the last of the blackface versions. The head was now 8½"x22"x9½", and the cabinet stayed the same.

Fender under CBS was going through a major transition in the design department, and their first change in the bias circuitry of the Bassman (model AA568) was a disaster! Resistors off the cathode at each power tube lifted the cathode above ground, electrically, and two capacitors were added to the bias circuit. The staff at the factory, the dealers, and the musicians all reportedly complained about the sound of the new amp, and CBS returned the cathodes to ground. The caps stayed, however, and a number of bias circuits were later experimented with.

By fall of '68 the 2x12 cabinet was replaced by a 2x15 setup in the same-size box. In mid to late '69 the aluminum trim was discarded along with the disastrous first-series solid-state amps (see page 144 for the Solid-State Bassman). This

Super Bassman II, '69–'71. *The "II" meant two bottoms; the "I" came with a single bottom. The first big bass amp.*

Bassman 10 ('72), '72–'82. *Four 10s in a sealed-back cabinet.*

standard silverface style would last into the eighties basically unchanged. A new amp, the Super Bassman (and Super Bassman II), was released around the time the aluminum trim was dropped, but these amps were made in both styles. This was Fender's most expensive tube amp, costing almost 50% more for a Super Bassman II with JBLs than for a Dual Showman Reverb with JBLs. The head was larger than that of the standard Bassman, putting out 100 watts. The box measured 13¼"x26"x9½". The 2x15 bottom was also bigger than that of the regular Bassman: 45½"x30"x11½". The difference between the I and II was the number of bottoms purchased with the head, which just said "Super Bassman." The regular 50W Bassman was still available, but had become an anachronism of sorts among the plethora of big, powerful bass amps in the late sixties.

Upon the release of the monstrous 400 PS Bass amp, the Super Bassman was temporarily discontinued; the Bassman became the Bassman 50 by 1972, with a new 2x15 cabinet (30"x28"x12") that had the speakers in the corners. The new 50W combo Bassman 10 was added to the line around this time and featured four 10s. The back was sealed, making it very unlike the tweed 4x10s. The first channel had Bass and Normal inputs, a Deep switch, and Volume, Treble, and Bass controls. The second channel had Bass and Studio inputs, a Bright switch, and Volume, Treble, Middle, and Bass controls. Master Volume, new for Fender, was included on this new amp, but don't expect much out of this feature.

These amps were seen in the 1972 catalog, along with the new Bassman 100, essentially a Super Bassman in a different package. The bottom measured 40¼"x30¼" across the front and was a whopping 17" deep, housing four 12s. The head measured 13¼"x26"x9½". Again, 100 watts, two channels with two parallel inputs in each, and Volume, Treble, Bass in the Bass Instrument channel, and Middle added to the Normal channel. Again, a new bass amp with a Master Volume. Soon even the Bassman 50 would be equipped with one.

By the mid seventies Fender discontinued its practice of bolting the head to the speaker bottom, and the logos became tail-less. By the end of '77 the names were changed to coincide with the upgrade in power claimed by the company: the 100 became the 135, and the 50 became the 70. In the eighties, blackface became first an option and then standard issue. A solid-state Bassman Compact (see page 150) was available for a short time early in the decade. In '82 the short-lived Bassman 20 was added, somewhere between the discontinued Musicmaster Bass and Bassman 10 amps. The 1x15 combo was powered by a two-6V6/two-7025, 18W amplifier with Volume, Treble, Middle, and Bass controls and a single input. The year 1983 saw the last of the tube Bassmans, which were phased out during the "II" era.

When the new company decided to market a reissue amp to go with their immensely popular reissue guitars, they chose the tweed Bassman over any number of excellent amps actually designed for the guitar. The 4x10 Bassman reissue, released in 1990, floored the public and led to a whole series of reissue amplifiers (see page 209). Its continued success shows the timelessness of the original design.

Bassman 100 ('72), '72–'77. Replaced the Super Bassman. The 4x12 cabinet is a gem. Became the Bassman 135, lasted till '83.

Bassman 20 ('82), '82–'83. A 1x15 combo for recording, bedroom use, or acoustic accompaniment. Designed for tone and portability.

TWIN
1952–1985
1987–Present

Twin
Twin Reverb
Twin Reverb II
The Twin

Apple Pie...Fender Twin...some things you can't go wrong with. Some recipes have basic ingredients, others are more elaborate. An old Twin is like Grandma's pie, hot out of the oven, simple and delicious. The silverface Twin Reverb, a dependable workhorse for years, is like a piece of diner pie, sitting out all day, nothin' else on the menu, but if you're hungry, it'll hit the spot. And a 1995 "Twin-Amp," with its 40-plus years of advanced technology, is for the more adventurous—nouvelle cuisine, so to speak.

Unveiled at the National Association of Music Merchants (NAMM) show in 1952, the unnamed "hi fidelity" amplifier was touted as a major breakthrough in music reproduction. Besides the new *wide-panel* tweed cabinet (20"x26¼"x10") with chrome control panel, On/Off switch, fuse holder, pilot light, four inputs (two Mic, two Instrument), and two Volume controls, the Twin was described in the promotional flyer of early '53 as having twin 12" heavy-duty Jensen "Concert Series" speakers, heavy-duty dual output tubes, a standby switch, a spare speaker jack, and separate Bass and Treble tone controls. The dual tone controls were a major breakthrough compared to the single tone roll-off pot on all the preceding models. A letter accompanying the flyer claimed the "Twin 12 Amplifier" would be "without a doubt the finest Amplifier available anywhere for musical instrument amplification." All the new wide-panel amps were "available for immediate delivery, with the exception of the Twin 12 Amplifier which should be ready within two to three weeks." The production model was advertised in *International Musician* with the name "Twin 12 Artist's Model Amp."

Articles in the trade magazines that summer described the tone circuit and improved power for both the Twin and the new Bandmaster (unspecified, but approximately 25 watts). Four identical inputs each went directly to one half of a 6SC7 preamp tube (two twin triode tubes in all). These tubes were grid-resistor biased and were followed by Mic Volume

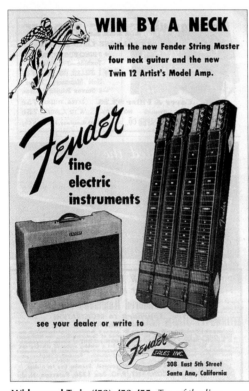

Wide-panel Twin ('53), '53–'55. *Top of the line.*

(slightly brighter) and Instrument Volume controls. The two channels were summed into a 6J5, a metal single-triode tube also found in some of the K&Fs. A 6SC7 *paraphase* inverter, two 6L6 power tubes, and a 5U4G rectifier completed the circuit.

The first in a series of continual upgrades (model 5D8) included changes in a number of areas. The input tubes changed to cathode-biased 12AY7s. The 6J5 single triode was replaced with a 12AY7 twin triode, direct-coupled, with the second stage a cathode follower. This made the tone controls more responsive. A Presence control on the negative-feedback loop and a change to a *self-balancing paraphase* inverter (12AX7) were other improvements. The 5U4 rectifier wasn't changed immediately, but would be replaced by two 5Y3GTs.

Nineteen fifty-five brought major changes both electronically and cosmetically. A somewhat smaller cabinet (20½"x24"x10½") featured the new *narrow-panel* style, as well as a new alignment of the speakers (in the bottom right and top left corners). Acoustically transparent grille cloth of contrasting shades of brown replaced the earlier monochromatic linen. The big change occurred inside, with the new amp (model 5E8) reportedly upgraded to 50 watts. A number of changes were responsible for the increase, including the replacement of the two 5Y3s with 5U4Gs and the addition of a choke in the DC filter section. A new power transformer with a fixed-bias tap for the power tubes made the two 6L6s more productive than the original cathode-biasing arrangement. The phase inverter circuit changed to a *split-load* style, using only half of the 12AX7 twin triode. The freed-up other half was used as another gain stage. This helped compensate for the small loss from having a second negative-feedback loop added from just before the tone circuit back to the grid of the second stage. Similar models for '56 and '57 featured the new ground switch and the exclusion of the treble circuit from the negative-feedback loop (model 5E8-A).

Narrow-panel Twin ('56), '55–'60. *New circuitry boosted the power to approximately 50 watts. Note the speaker configuration, c. '55–'58.*

FENDER AMPS: THE FIRST FIFTY YEARS

Affectionately known as the "little deaf girl" ad.

Narrow-panel Twin ('59), '58–'60. "Big-box" version with lots o' power, from four 5881s. Speakers returned to side-by-side layout. Separate Bass, Treble, Middle, and Presence controls.

Yet another big increase in power was unleashed the following year, a new model that boasted four power tubes. What a different amp this is, today being considered the *crème de la crème*. No power ratings were given in the '58 catalog, and the text didn't make a note of the increase, except in the list of tubes: one 12AY7 (preamp), two 12AX7s (one preamp, one *long-tailed* phase inverter), four 5881s (power), and one GZ34 (rectifier). This list describes model 5F8-A. A short-lived earlier version (5F8) used an 83 mercury-vapor rectifier tube, as seen for a short time on 4x10 Bassmans. The picture in the '58 catalog was a repeat from the previous year, showing four small tubes instead of three, and only five knobs. The actual amps had six knobs: Normal and Bright Volumes, Bass, Treble, Presence, and the new "Middle" control (built into both the Twin and the Bassman). The *Down Beat* insert from July '58 mentions this upgrade, but shows only the front of the amp. The '59 *Down Beat* insert, however, clearly shows a six-knob panel. These are often referred to today as 80-watt or 100-watt tweed Twins, and in many situations they were more amp than the Jensen P12-N speakers could handle. Speaking of speakers, they were again mounted side by side in a larger cabinet, and that was more than the old leather handle could handle. It was replaced, c. late '58, with a brown plastic model, as seen on the last tweed Bassmans and Bandmasters, as well as the large amps of the early sixties. Some (at least one) of the last top-mounted, chrome-panel Twins, c. late '59–early '60, were covered in the new brown Tolex as seen on the Vibrasonic and Concert.

As important an amp as the Twin was for Fender, it ended up in limbo for a short period. Superseded at the top of the line by the Lansing-equipped Vibrasonic and bypassed by the R&D department, who were perfecting the Lansing-equipped piggyback Showman prototypes, the speaker-eating Twin was not originally part of the new "Professional" series of brown-Tolex-covered, front-control-panel, "Harmonic Vibrato"-equipped, two-6L6 amps. These were described in an article in the April '60 issue of *Musical Merchandise Review*, which listed the Vibrasonic, Concert, Super, Pro, and Bandmaster amps. The missing brown Twin showed its face two months later in the June '60

Down Beat insert, along with the other Professional amps. All were pictured with the control layout of Bass, Treble, Volume, Speed, Intensity, and Presence, including a small-box brown Tolex Twin with its logo running diagonally. Although a shot of the backside of the amp was not included, it appears that the speakers would have to have been mounted in opposite corners, as on the '55–'58 models. (Two reliable sources swear to having seen this configuration.) The description of the amp included the tubes used: five 7025s and—"Drum roll, please..."—*two* 6L6 power tubes! The 1960 full-line catalog listed the Twin in its two-page spread on the Professional amps, and a brief description (in a rewrite of the '58 catalog) appeared on the page with the premier Fenders of the day: the Jazzmaster, the Stratocaster, and the Vibrasonic. Here, too, it was listed as having *two* 6L6s. No mention was made of coverings, controls, or tremolo, and the amp, unfortunately, was not pictured.

All the 1960 Professional amps have schematics showing five 7025s, two 6L6s, and the volume controls in the middle. The schematic numbers are 6G4 (Super), 6G5 (Pro), 6G7 (Bandmaster), 6G12 (Concert), and 5G13 (Vibrasonic, released in '59). There doesn't appear to be a schematic for the early-to-mid-'60 Twin. The earliest Twin schematic of the sixties (6G8) shows *six* 7025s, four 6L6s, and Volume, Treble, Bass, etc. for the controls. The schematics for the other Professional amps that show six 7025s and Volume, Treble, Bass are numbered 6G4-A, 6G5-A, 6G7-A, 6G12-A, and 6G13-A, corresponding to amps from *late* '60 and on. Since all the amps from early '60 were about the same, it seems logical that Fender had a few control panels (at least one) made up with the "Twin-Amp" logo and installed them on the chassis of one of the other models. We're talking about a very short period here. The April *Musical Merchandise Review* article was probably turned in by late March, and by the time the *Down Beat* insert was turned over to the magazine (late May for the June issue?), the cabinet size had been changed to that of the late-'60-to-mid-'63 Twins (19"x27½"x10¼"). So you're looking at, tops, a two-month period for the small-box brown Tolex Twin. Fender made a few (again, at least one) brown Tolex Twin amps in the 19"x27½" cabinets with the speakers mounted side by side c. late '60–early '61 (see page 191). The control panels on

Transitional brown Tolex Twin ('60).

these have the "Twin-Amp" logo running horizontally and the control lineup of Volume, Treble, Bass. These would become the legendary white Tolex Twins in '61. Needless to say, the transitional brown Tolex Twins are very rare amps, but they do exist! (P.S.: The July '60 price list still listed all the amps but the Vibrasonic and Concert as having Linen Covering, the document obviously having not been updated.)

White Tolex Twin ('62), '61–'63. *Preceded by version with brown Tolex and tweed-era grille (see page 191). Model pictured fitted with maroon grille cloth ('61–'62). Later version had wheat grille ('62–'63). Powerful, like the last of the tweed Twins, plus the aural splendor of tremolo.*

Following the release of the four-5881 Showman in late 1960, the Twin received a new look: the same cabinet as the last brown Tolex models, but with white Tolex and maroon grille. Its two lost power tubes had returned (model 6G8), and this circuit was identical to the Showman (6G14) with the exception of the 4Ω output transformer. The controls read Volume, Treble, Bass on the "Normal" channel, and Volume, Treble, Bass, Speed, Intensity on the "Vibrato" channel, with Presence active on both. The midrange control was still gone. The '62 model was identical until late in the year, when it changed to wheat grille cloth. The brown "Fender" handle was replaced in early '63 with the black metal-reinforced style, and this final white-knobbed version officially lasted until the middle of '63 (a few were still coming out of the factory as late as December '63).

Then came the big change: Reverb, and a new black Tolex cabinet (20¼"x26⅜"x10½"). Two different pictures of early Twin Reverbs showed them with an unusual solid gray grille cloth. The first picture was in the *Down Beat* insert of September '63, and showed the old flat logo; the second was on the cover of that insert, with the new raised logo (albeit mounted somewhat out of kilter). A similar shot, in full color, appeared on the cover of the '63–'64 full-line catalog (see page 194). Inside that catalog was a color picture of the third and final variation, which would become standard production for the next four years. This amp was fitted with the classic silver sparkle grille and a properly positioned raised logo.

The Presence control was replaced with a Bright switch on each channel, and the Middle control was brought back following a three-year hiatus. The white knobs were replaced with black, skirted versions with numbers, and the control panel changed to the black numberless style. The tremolo circuit used two fewer tubes than the previous three-tube version by using a photoresistor to send pulses of the preamp signal to ground. The new Reverb section included a two-spring pan driven by a 12AT7 tube, with half a 12AX7 used for the recovery. The phase inverter tube was changed to a 12AT7—Fender's new choice for all their push-pull amps—and the preamp tubes remained 12AX7s. A courtesy AC outlet and an extra RCA jack for the Reverb Pedal were new additions to the back panel.

The blackface Twin Reverb speaks with a deep, rich voice, sonorous and confident. It is possibly the quintessential combo amp, having a wide tonal range, the sweet sound of spring reverb—as well as tremolo—and enough reserve power to fill most environments without being miked. This was most important in the sixties, becoming less of a consideration as the P.A. systems of the seventies progressed. But if you play with a hard-hitting drummer, being able to get a clean sound at a satisfactory stage volume can be difficult with 15–50-watt amps. This has made the Twin Reverb popular for years with country and jazz players, electric piano tinklers, and other non-rock musicians. Or turn it on 10, *à la* Ted Nugent, Johnny Thunders, et al.

The '68 Twins changed to the cheerful chrome-and-blue-sparkle attire with aluminum trim and a slightly shallower cabinet (9¾" deep). The amp itself was unchanged, but only for a short time. The engineers changed the bias section (model AC568) in an attempt to further stabilize the amp. Many of the changes were undone within a few years, but the Bias Balance control, which replaced the Bias Adjust pot, stayed on.

Twin Reverb ('76), '63–'82. *One of the most popular amps of all time. The blackface version ('63–'67) was followed by silverface with aluminum trim ('68–'69), silverface as pictured ('70–'81), and a return to blackface ('81–'82).*

The '72 catalog introduced the new Master Volume control, which was a good idea, but was only included on the large amps, which were basically designed to stay clean at any level, minimizing the effect. These amps work quite well as Twins, though, and can be found secondhand for less than the cost of installing new transformers. Casters had become standard equipment, and the cabinet settled into the dimensions it would keep until the end (20"x26"x10½"). In the mid seventies (by '76) the logo lost its tail, and another well-intentioned circuit upgrade was made: the Boost circuit. This was activated by a new pull-knob volume pot, and actually was a practical feature, though not as controllable as a true channel-switching system. A Hum Balance control was also added around this time. By the end of '77 the power was reportedly increased to 135 watts, which is really pushing the tubes, but gives a loud, clear sound. A Line/Recording out jack and an Output Tubes Matching circuit were added shortly thereafter.

The blackface look was offered as an option beginning in 1980 and became standard issue for 1982, the last year of the traditional Twin Reverb.

Twin Reverb II ('83), '83–'86. *A revved-up version, trading tremolo for overdrive.*

By the end of '82 the 105-watt Twin Reverb II was released, replacing the tremolo circuit with three-stage distortion and channel switching. An effects loop with rear-mounted level controls, the return of the Presence control, a pull-knob Mid Boost, and the placing of the On/Off and Standby switches on the front panel were the other obvious changes. Besides having a deeper cabinet (26⅛"x19⅞"x11⅜"), the amplifier section could be purchased separately in a package that should have been called "The Showman," except that the name had already been given to a new line of solid-state amps. The Twin II amps were offered at least into '86, though production had stopped at the end of CBS control.

The first series of new amps following the change in ownership included, of course, a new 100-watt 2x12 combo, definitively named "The Twin." A new look featured dark gray grille and red knobs. A traditional clean channel and an overdrive channel could be accessed in three ways: individually (for two instruments), in a channel-switching mode, or in parallel, which made every knob on the amp active. This allowed the simultaneous combining of clean and distorted sounds. "Great for the studio." A low-power (25 watts) switch, pull-knob boost on the tone controls, pull-knob Presence control with Notch Filter, and assignable Reverb gave the user an infinite variety of sounds. The back panel included a serious Bias Adjust and Balance control, selectable impedance, a three-level Effects Loop, and a Balanced Line Out. A choice of five custom-color coverings was available for the adventurous (cabinet size 22"x26"x12"). The red knobs and dark grille were replaced with black and silver, respectively, and the success of this amp kept it in the line after the rest of the initial amps from FMIC had been retired.

"The Twin" ('87), '87–'94. *The new Fender company put all their engineering know-how into this totally reworked extension of the powerful 2x12 combo. The red knobs would be replaced with black for the last few years.*

For the more traditional player, Fender began offering a reissue of the '65 Twin Reverb in '91, following almost 10 years of Twins without tremolo. The Reissue is very close to the original, both in looks and in circuit; but the replacement of most of the wires with a printed circuit board, and ¼" jacks for the reverb, should keep any of these from being passed off as originals.

As traditional as the Reissue is, the '95 "Twin-Amp" is equally modern. The back panel has On/Off, Standby, and High/Low Output (100 or 25 watts) switches, an Effects Level switch (–13 to +2 dB), Effects Send and Return jacks, Effects Select (channel 1, 2, or both) and Reverb Select (ditto) switches, Preamp Out and Power Amp In jacks, a two-pot Bias Adjust and Balance circuit with test points allowing for adjustment without the removal of the chassis, a jack for the foot switch that controls Channel Select, Gain Select, and Reverb on/off, a Balanced Line Out (+3 dB) XLR jack, a

Reissue '65 Twin Reverb, '91–. *Following the success of the Reissue '59 Bassman and brown Vibroverb, Fender added yet another classic to their Reissue line.*

FENDER AMPS: THE FIRST FIFTY YEARS

"Twin-Amp," '95–. At first glance this amp may look like a '65 Twin Reverb, but it marks the latest step in the technological progression started with the Twin Reverb II.

Main Speaker out as well as Series and Parallel Extension Speaker jacks, and an Impedance Selector. Whew! They had to turn the serial number sideways for it to fit!

The front panel is much simpler, with a pilot light, the logo, two input jacks, and 14 knobs! Actually, the knobs are straightforward, with a pull-knob Bright switch being the only feasible distraction. Two switchable channels each have Treble, Bass, and Mid tone controls, with a master Presence control. A variety of Gain and Volume controls work in conjunction with the foot switch to give three distinct sounds or levels. A Mix control for the effects loop (see the '93 Concert) and a master Reverb control complete the arsenal. Between this amp and the Reissue model, Fender should continue its tradition of the Twin as first choice for touring professionals, rehearsal studios, and rental businesses. Old Reliable.

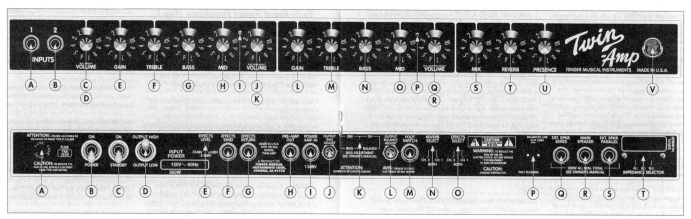

'95 Twin-Amp.

88

BANDMASTER
1953-1980

Bandmaster
Bandmaster Reverb

Looking back on the release of the two-6L6 1x15 Bandmaster in mid 1953, one must ask the question, "Why?" Fender already had the two-6L6 1x15 Pro, not to mention the two-6L6 1x15 Bassman. The trio's *wide-panel* cabinets even shared the same dimensions (20"x22"x10"), as did the new 1x15 Extension Speaker. So what was special about the Bandmaster? The "New Fender Tone Control Circuit," which was a great improvement over the simple treble roll-off Tone knob of early amps. Like the new Twin, the Bandmaster had both Treble and Bass controls.

The Bandmaster was not included in the foldout brochure of early 1953 that introduced the Twin and the wide-panel cabinet. An accompanying letter made no mention of the upcoming availability of the Bandmaster, but mentioned that the Twin 12 Amplifier "should be ready [for delivery] within two to three weeks." Ads pushing the Bandmaster and Twin amps began running in the trade magazines by June '53, and articles from that time described the new amps as having the same power and tone controls. The $229.50 Bandmaster (model 5C7) cost $20 less than the Twin (model 5C8) and $30 more than the previously top-of-the line Pro. Prototype Bandmasters with the old *TV-front* cabinets are rumored to exist but were never catalogued. However, having a lower model number (7) than the Twin (8) implies Fender was working on the design in '52, so prototypes from the TV-front era are possible, though highly improbable.

Wide-panel Bandmaster ('54), '53–'55. *Entering the line at the #2 spot, just below the Twin, the 1x15 Bandmaster was essentially a Pro with separate Bass and Treble controls.*

The first production models (e.g., serial no. 0084, dated June '53) had three inputs (two Instrument and one Mic) and four controls (Instrument Volume, Mic Volume, Treble, and Bass). On/Off and Standby switches, a pilot light, a fuse holder, and a stenciled-on logo completed the top panel, and an extension speaker jack was mounted on the bottom of the chassis. Two cathode-biased 6L6G power tubes followed a

Narrow-panel Bandmaster ('56), '55–'60. *The Bandmaster came into its own with the release of the 3x10 edition.*

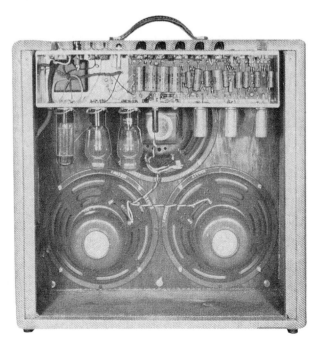

'56 Bandmaster with back panels removed.

conventional *paraphase* inverter (6SC7). Two more 6SC7s were used in the preamp section, and a 5U4G rectifier powered the amp. The tube chart on this early example had the original name and model number inked over, with "Bandmaster" and "5C7" hand-written above. This tube chart must have been from a Super or a Pro, the other amps from this era with the same tube configuration, although the actual circuit of the Bandmaster was quite different. The Jensen P15-N speaker in this unit shows a date of the twentieth week of 1953.

In 1954 a revised model (5D7; e.g., serial no. 5491, dated September '54) replaced the 6SC7s with two 12AY7s and a 12AX7 *self-balancing paraphase* inverter. The '54 catalog specified four controls, although it pictured a three-knobbed amplifier. This was the same photograph as the Pro, which was also used for the catalog shot of the two-knobbed 1x15 Bassman. From the front the three amps were identical, so separate photos must not have seemed necessary. The 1x15 Bandmaster was offered on price lists until July of '55.

Although there was no Bandmaster in the *Down Beat* insert of July '55, implying a temporary halt in production, a new *narrow-panel* version with three 10" Jensens (model 5E7) replaced the old model for the '55–'56 full-line catalog. Conservatively rated at 26 watts, this is the amp most players think of today as a tweed Bandmaster. A Presence control and a Ground Switch were new additions to the control panel. A new logo plate with "Fender Bandmaster" in script set off the front of the restyled cabinet (21¼"x22½"x10½"), as did the new acoustically transparent grille cloth. The preamp section included a 12AY7 (one half for the two Bright inputs, the other half for the two Normal) and three stages of 12AX7 preamp common to all four inputs, including a direct-coupled cathode follower just before the tone circuit. Other new designs included the employment of negative feedback in *two* sections, a fixed bias supply for the two 6L6G power tubes, and the use of only one half of a 12AX7 for the new *phase splitter* (split-load inverter). A 5U4G rectifier was still standard, but the DC filter section now included a choke to supplement the filter caps. The '55 Bandmaster was considerably more advanced electronically than the versions released just two years prior. Two 5881 power tubes replaced the 6L6s c. '57; otherwise, little was changed until 1960. Models from '59 still used the 5E7 circuit. The last of these were fitted with the new brown plastic handle.

As part of the new Professional series, announced in April '60, the Bandmaster (model 6G7) joined the recently released Vibrasonic and Concert models, as well as upgraded versions of the Pro and Super. The plastic handle, cylindrical knobs, Tolex covering, and front-mounted control panel were all various shades of brown, and the look was highlighted by a shiny new flat logo. Two discrete channels had two inputs each, with controls as follows (*L to R*): Bass, Treble, Volume (Normal channel), and Bass, Treble, Volume, Speed, Intensity (Vibrato channel). A master Presence control worked on both channels. Apparently, a few remaining cabinets with top-mounted chrome panels were covered in the new brown Tolex.

The preamp for the Vibrato channel used both sides of a 7025, surrounding the volume and tone circuitry, keeping it from interacting with the controls of the Normal channel. A four-stage oscillator/crossover tremolo circuit using two more 7025s operated on the Vibrato channel only, and was vastly more complex, both electronically and harmonically, than earlier and later tremolos (see Vibrasonic). The Normal channel also used both sides of a 7025, and the two channels joined at the *long-tailed* phase inverter (7025). Two 6L6GC power tubes and a solid-state rectifier using six silicon diodes completed this all-new circuit.

Brown Tolex Bandmaster, '60. *The 3x10 design carried over into the short-lived model with brown Tolex and front control panel.*

Towards the end of 1960, the Bandmaster's controls changed to Volume, Treble, Bass, as did those of the rest of the Professional series. A refined five-stage tremolo circuit employed two 12AX7s (one side of the second tube was left unused) and a 7025. The Presence circuit was changed slightly, as were the inputs, but the power section was unchanged, save for the return of 5881 power tubes. This is the amplifier that would end up as the piggyback Bandmaster.

Announced by February '61 as a companion to the Showman and Bassman, the new piggyback Bandmaster (model 6G7-A) sported the white Tolex/maroon grille/white knob look. A head measuring 8"x24"x9" and a ported 1x12 speaker cabinet (30"x21"x11½") replaced the old 3x10 combo style (which would return in '93 on the Vibro King). A 2x12 sealed cabinet (32"x21"x11½") replaced the 1x12 bottom for '62. Changing cosmetics in the standard order for models of the early sixties, the Bandmaster switched to a wheat grille later that year. A black reinforced handle replaced the brown plastic version by mid '63.

Piggyback Bandmaster with white knobs ('61), '61–'63. *The first model was equipped with one 12" speaker in a ported cabinet, followed by a 2x12 sealed cabinet. These two amps were appointed with maroon grille cloth. In late '62 the cloth changed to wheat-colored. Featuring all-tube tremolo.*

The '63–'64 catalog and the *Down Beat* insert of September '63 show this style of cabinet, but with the new blackface, black-knobbed control panel. This amp used a new circuit design (model AA763) that replaced the Presence control with Bright switches and implemented the new photoresistor tremolo circuit. This tremolo circuit employed two fewer tubes than the earlier version by sending the pulses of the Vibrato channel signal to ground as opposed to oscillating the grid voltage. A 12AT7 became the standard phase inverter tube at this time, and a bias adjust pot was added to the bias section of the power tubes. This model would be the basis of all the Bandmasters that followed, making the last of the white Bandmasters more akin to the later black Tolex blackface ('64–'67) and silverface ('68–'75) amps than to the models from '61 through mid '63. Not that one is better sounding (or looking) than the other, but there is a difference.

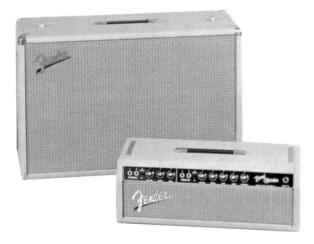

Blackface Bandmaster ('63), '63–'67. *A new circuit, with Bright switches replacing Presence control and remodeled tremolo, make the last of the white Bandmasters noticeably different from preceding models. The wheat grille would change to gold in late '63, and the white Tolex would be replaced with black by mid '64. The cabinet was enlarged greatly in '67.*

The flat logo was replaced with the raised version, and gold sparkle grille cloth joined a smoother white Tolex by late '63. In mid '64 the black Tolex and silver sparkle grille stabilized the cosmetics until the silverface-with-aluminum-trim version of '68. The original small 2x12 cabinet was replaced on the last blackface models with a version 29½" wide by 40" tall (c. '67), which had the speakers mounted one above the other, instead of side by side. With the cabinet in its upright position, the T-nuts were on the "top," the handles were on one side, and casters were on the other. Needless to say, the tilt-back legs were removed at this time. It seems probable that the enlarged cabinet was designed more to make the amp look impressive than for sonic reasons.

Along with the change to silverface with aluminum trim and blue grille, and following the discontinuation of the tube Reverb Unit, another Bandmaster was added to the line. For $40 extra, the Bandmaster Reverb (model AA768) offered the benefits of a sealed-bottom amp with the classic sound of reverb. Why Fender waited until the summer of '68 to do this instead of '63 when surf music was still popular is a mystery, unless you consider that the price of the essential Reverb Unit was $129 in '63. These early models also were referred to as the TFL5005.

The circuit for the new amp, as well as for the revised Bandmaster (model AA568) included a major change in the bias section of the power amp. An overwhelmingly negative reaction from everyone but the designers led to the reversion of a few of the changes (model AA1069); but much of the "improved" circuitry remained, including the use of the old bias adjust pot for balancing the outputs of the power tubes. The aluminum trim was deleted from the cabinets around this time, and in 1970 a grounded AC plug and a three-position Ground Switch were added to the rear panel.

The Bandmaster was discontinued in '74, but the Reverb version lasted into '80. Mid-seventies additions included a Master Volume with pull-knob Boost, a Middle control in the Vibrato channel, a Hum Balance control on the rear panel, and a squat new cabinet, still 2x12 but with speakers in the top right and bottom left corners (30"x28"x12"). Removable casters were standard, but the pull-out straps on the head (used to bolt it to the cabinet) were removed. All these changes took place by the release of the '76 catalog, as did the change to the tail-less logo with the "®" symbol. The model was last sold in '80. An interesting listing on the October '88 price list showed a "Bandmaster Head," which was a match to the Super 60 combo. Were any of these actually made? By the end of the year it became the "Super 60 Top."

Silverface Bandmaster and Bandmaster Reverb ('76), '68–'80.
Silver control panel, blue sparkle grille cloth, and aluminum trim separated the '68 model from the blackface amps, but at first their internal workings were unchanged. A deluxe version, with reverb, was added to the line later in the year. These two amps would lose the aluminum trim c. '70. The reverb-less model would be retired in '74, but the Bandmaster Reverb would last another six years. The amp pictured here, c. '76, featured a new, squat cabinet.

TREMOLUX
1955–1966

Tremolo, a.k.a. amplitude modulation, is a very pleasing effect and probably the first built into an amp, starting in the late forties with Danelectro, Gibson, and Premier. So when Fender introduced its new Tremolux amp in the summer of '55, it was not a major event. To those who today worship "that sound," still available on certain new Fenders, the Tremolux was a very important amp. Basically a Deluxe with tremolo, the first Tremolux amps (model 5E9) shared the same cabinet as the Deluxe (16¾"x20"x9½"). The electronics were put into the cabinet of the Pro starting in 1956 (20"x22"x10"). Bright and Normal Volumes, Tone, Speed, and Depth were the controls, with an On/Off switch, a pilot light, and a fuse holder completing the panel. A wedge-shaped foot switch that plugged into the bottom of the chassis with a ¼" phone plug allowed the tremolo to be brought in or out for special effects.

Narrow-panel Tremolux ('55), '55–'60. *Based on a narrow-panel Deluxe, the Tremolux was Fender's first amp with a built-in effect: tremolo.*

The early tweed Tremolux's two 6V6 power tubes were cathode biased, unlike the fixed bias of all the tremolo amps that followed. Its method of connecting the oscillator (12AX7) to the cathode bias of the *paraphase* inverter tube (12AX7) was also unlike any that would follow. A 5Y3 rectifier and a 12AY7 for the preamp section were the other tubes.

In '57 the power rating was reportedly increased from 15 to 18 watts—an unnoticeable change to the human ear. This version (model 5G9) would be Fender's most powerful tweed amp with tremolo. The tubes were the same, except for a 5U4GB replacing the 5Y3. An extra filter cap and a choke beefed up the DC supply. A new power transformer with a fixed-bias tap for the power tubes allowed the Depth pot to be connected to the bias supply, the oscillator, and the grid resistors of the power tubes. Varying the bias voltage made the volume go up and down in sync with the oscillator. With the removal of the Depth control from the cathode bias of the phase inverter tube, a new *long-tailed* phase inverter was employed.

Basically unchanged through the late fifties, the amp underwent a major transformation in 1961. The Tremolux went from a 1x12 tweed combo with two 6V6s to a 1x10 piggyback (bottom: 17½"x27"x11½") covered in white Tolex and for a short time equipped with two 6BQ5 power tubes (model 6G9). Two Bright inputs were on the left side of the control panel, with knobs for Volume, Treble, and Bass. The Normal channel also had knobs for Volume, Treble, and Bass, with Speed and Intensity to the right. Tremolo, which was now available on every amp but the Bassman and the Champ, was about the only thing the old and new Tremoluxes had in common. A GZ34 rectifier replaced the tweed's 5U4, and a pair of 7025s replaced the single 12AY7, isolating the Bright and Normal channels from each other. Unlike the bigger amps, the tremolo worked on both channels. The tremolo (12AX7) and phase inverter (12AX7) tubes remained unchanged.

Two 6L6 power tubes quickly replaced the 6BQ5s (model 6G9-A), but the power of these Tremolux amps was always considerably less than what Fender regularly squeezed from 6L6s. From this model on, the Normal channel would be on the far left. The tone ring used to mount the speaker (see Showman) was discontinued in '62, the single 10" being replaced with two 10" speakers (model 6G9-B) in separate enclosed air spaces. (Cabinet dimensions remained unchanged.) Later that year the maroon grille cloth was changed to wheat colored. A new black handle by mid '63 was the last change to the white-knobbed Tremolux.

The '63–'64 catalog showed an interesting amp, having the new black-skirted knobs placed over the numbers on the old brown control panel. There were no Bright switches, and the Tremolo was for both channels. Obviously the numbered knobs were added just for the photo shoot, but soon they and a numberless blackface control panel were standard. This amp (model AA763) was considerably different from the white-knobbed models. The tremolo was the new style that used a

Piggyback Tremolux with white Tolex and white knobs ('62), '61–'63. *The first model came with maroon grille cloth and a single 10" speaker, which soon changed to two 12s. In late '62, wheat grille cloth was added. The tremolo worked on both channels.*

Blackface Tremolux with white Tolex ('63), '63–'64. *A new circuit included tremolo that worked on the Vibrato channel only and a Bright switch. The wheat grille cloth was replaced with gold in late '63, and the white Tolex was replaced with black in mid '64. The amp used for this photo, from the '63 catalog, is actually a white-knobbed brown-panel version fitted with black knobs.*

Blackface Tremolux with black Tolex ('64), '64–'66. *The final version.*

12AX7 and a photoresistor to dump the signal of the Vibrato channel to ground. The tremolo no longer affected the Normal channel. A 12AT7 phase inverter replaced the 12AX7, and a 10kΩ bias adjust pot was added. The Bright switches on the front panel are the giveaway features in identifying this circuit. The flat logo was replaced with a new raised version and the Tolex became a smoother texture in late '63. The last change to the white Tremolux was the addition of gold sparkle grille cloth by the end of the year. In mid '64, black Tolex with silver grille cloth became standard; otherwise the amp was unchanged. It was yet another short-lived variation, as the Tremolux was discontinued in the summer of '66. The phase-out coincided with that of the Showman 12, as Fender released its first solid-state amps.

HARVARD
1955–1961

The Harvard amp joined Fender's growing line in late 1955, either as the company's smallest amp for professional use or as a super practice amp. With a single 10" speaker and two 6V6 power tubes producing 10 watts, the Harvard filled a gap between the 1x12 15-watt Deluxe and the 1x8 4½-watt Princeton. Like the Princeton, the Harvard featured Volume and Tone controls, a fuse holder, a pilot light, and an On/Off switch. The dimensions of the *narrow-panel* cabinet matched those of the big-boxed, narrow-panel Princeton. Price lists starting with November '55 through February '61, and the eight-page catalogs from '56 to '60, specified a 10" speaker; but the full-line catalogs from '56 to '60 specified an 8". At least one version of the single-page spec sheet from the dealer notebook shows "8" hand-written over the top of "10." So what size speaker was in the real thing? It appears both were used at different times.

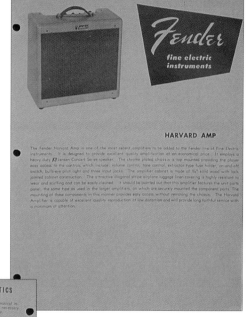

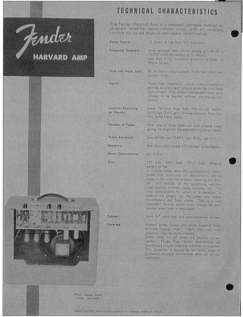

Harvard (c. '56), '55–'61. *A down-sized Deluxe aimed at the student market, the Harvard had a smaller speaker and only one channel. It was, however, equipped with two 6V6 power tubes, like the Deluxe. (Courtesy Hardtke Archives)*

A 12AX7 performed double duty, with one side as the second preamp stage and the other half as a *phase splitter* (split-load inverter). This had become standard Fender practice in '55, as had the fixed bias for the power tubes. A 5Y3GT rectifier was also standard issue on the smaller amps. The first preamp stage was a 6AV6 single triode, and this was *not* standard Fender practice at the time. Previously, the thrifty designers used *twin* triodes almost exclusively in the pre-amp and phase inverter circuits. The mate to the Harvard, the Vibrolux, was basically the same amp except for a 12AX7 in place of the 6AV6, with one half doing the 6AV6's job and the other used for the tremolo. On later amps needing a single preamp stage, Fender would use a 12AX7 and hook up only one side, a more practical approach than trying to find a 6AV6. A lower-gain version, the 6AT6, replaced the 6AV6 in '56 (model 5F10) and was used on the majority of the production models. (Is this a hint, Harvard owners?)

The Harvard was discontinued when the Princeton was upgraded to two 6V6s, a 10" speaker, tremolo, and brown Tolex in '61. An infamous schematic from this period for a Harvard with a single 6V6 (model 6G10) was certainly not for a production model, and is actually the same as the late-fifties Princeton (model 5F2A). Perhaps someone at the company considered a model between the new Princeton and the bottom-of-the-line Champ (which was still tweed and with just a volume control). The brown Princeton, a great deal at $89.95, and the Champ, at $62.50, were too close in price to justify a model in between. It was also not a Fender practice to do anything but upgrade their models, so it seems doubtful that Fender would even have bothered to make a prototype of this amp.

The Harvard name was revived for a short time from '81 to '85 and bears the dubious honor of marking Fender's reintroduction into solid-state guitar amps (see page 150).

VIBROLUX
1956–1982
1995–Present

Vibrolux
Vibrolux Reverb
"Custom" Vibrolux Reverb

Released in 1956 as a counterpart to the Harvard, the Vibrolux shared a number of its features: one 10" Jensen speaker and two 6V6 power tubes producing 10 watts. The tweed-covered Vibrolux also featured tremolo and was fitted in the same *narrow-panel* cabinet as the big-boxed Deluxe (16¾"x20"x9½"). It joined the tweed Tremolux as one of Fender's only two amps with tremolo—until the release of the brown Tolex Vibrasonic in '59. The schematic for the amp (model 5E11) is virtually identical to the Harvard: 5Y3 rectifier, two 6V6 power tubes, half a 12AX7 for the phase splitter and the other half as a second gain stage. Instead of the Harvard's 6AV6 single-triode preamp tube, the Vibrolux used a second 12AX7, with one half for the first preamp stage and the other half as an oscillator for the tremolo. This tremolo circuit was similar to the second Tremolux design, varying the fixed bias on the power tubes. A second variation (model 5F11) used slightly different values of capacitors in the oscillator circuit; otherwise the amps were identical. The control panel included three inputs, Volume, Tone, Speed, and Depth controls, a pilot light, an On/Off switch, and the fuse holder (¾ amp). Speaker and tremolo foot switch jacks were mounted on the bottom of the chassis.

In '61 the new combination of brown Tolex and wheat grille (18"x23"x9") dressed up a reworked Vibrolux (model 6G11), with two 6L6 power tubes and a 12" speaker. These amps did not develop as much power as many of the amps with two 6L6s, sharing an identical chassis with the low-power two-6L6 Tremolux of the early sixties. This variation would last until the change to blackface occurred in '63.

Two channels, Normal and Bright, each had Volume, Treble, and Bass controls and two inputs. Each used two sides of a 7025. Both sides of a 12AX7 were used for the upgraded *long-tailed* phase inverter, and both sides of another 12AX7 were used for the upgraded tremolo oscillator. The Speed and Intensity (instead of Depth) controls worked on both channels, still varying the fixed bias voltage on the power tubes. A GZ34 rectifier replaced the 5Y3 of the tweed

Tweed Vibrolux ('59), '56–'61. *Essentially a Harvard with the addition of tremolo.*

Brown Tolex Vibrolux ('61), '61–'63. *Upgraded to a pair of 6L6s, a 12" speaker, and two channels.*

Vibrolux. The On/Off switch was located on the back panel, and a Ground switch was added. An extension speaker jack was another new addition.

The blackface version, released in the summer of '63, sported a number of changes, including more power, black Tolex, silver sparkle grille (still with no logo), black numbered knobs, and, of course, a black numberless control panel. Circuit changes were similar to those of all the new blackface amps: a Bright switch for each channel, a 12AT7 phase inverter, and the new tremolo circuit. This used a photoresistor to open a path to *ground* for the preamp signal of the Vibrato channel. It also allowed the Normal channel to operate independently of the tremolo. An interesting typo on the control panel identified both channels as Normal, making this amp a young Republican's dream.

Blackface Vibrolux ('64), '63–'64. *Black Tolex, knobs, and control panel replaced the brown color scheme for this short-lived model. Two channels: Normal and Normal!?*

The blackface version lasted until the fall of '64, when it was replaced with the 2x10 Vibrolux Reverb. Fender would go without a powerful 1x12 combo, today a very popular match-up, for more than 15 years. It's a pity they didn't add reverb to the 1x12 two-6L6 blackface combo back then; it would today probably be the most desirable amp they made. The Vibrolux Reverb (model AA964) had two 10" Jensens and appeared more like an early sixties Super amp than the Vibrolux amp it replaced. Internally, however, the rectifier, power, phase inverter, tremolo, and preamp sections were very similar to the preceding Vibrolux. The addition of reverb in the Vibrato channel required two more tubes—a 12AT7 driver and a 7025 for the recovery section. A control for the Reverb (along with the proper labeling of the Normal and Vibrato inputs) was the only change to the front panel. Fender had a winning combination with this amp and kept it in the line, basically unaltered except for the standard cosmetic changes, until 1982, when all the old-style tremolo amps were retired.

Vibrolux Reverb ('65), '64–82. *A new 2x10 cabinet accompanied the addition of reverb. The blackface version lasted into '67. A silverface version with "blue" grille but similar circuitry ran from '68 until '81; the amps from '68 to '69 came with aluminum trim. In '81 the panel changed back to black. This version was the last of the Vibrolux Reverbs, being retired in '82.*

Silverface with aluminum trim replaced the blackface, and soon the bias section was changed to the balance-pot style, with the cathodes of the 6L6s lifted above ground, and a pair of small caps between the grids and ground. The last two upgrades were soon removed. The cabinet changed to 18½"x24¾"x9½", and a three-position ground switch was added c. 1970. A new logo was added in the mid seventies, and a Boost circuit activated by a pull-knob Volume pot was the only major change during that decade. The option of old-style blackface was offered in '80, and the following year it became standard issue on the amp. With the release of the "II" series in '82, Fender made the unwise decision to retire the Vibrolux Reverb. (After bouncing between Super Reverbs, which seemed too clean at the desired volume, and Deluxe Reverbs, which seemed too dirty, the author purchased a '66 Vibrolux Reverb in 1981 and found the amp of his dreams. Nearly 15 years and thousands of sets later, it still gets used regularly for gigs, rehearsals, and recordings. Is that a recommendation?)

A new "Custom" Vibrolux Reverb was released in early '95, sporting white Tolex, wheat grille cloth, and white knobs. The flat logo and black handle make this amp a reissue of the early-to-mid-'63 piggybacks and Twin, stylistically. Not being a reissue of any particular model, the new Vibrolux was released as part of the "New Vintage" Series Custom amps, along with the blackface Vibrasonic and the tweed Reverb Unit. The tremolo and reverb circuits operate differently from those of the original, working on both channels. The Vibrolux and Reissue Vibroverb are similar conceptually, with the Vibrolux having blue Alnico speakers instead of the ceramics found on the Vibroverb. If only because it is reminiscent of the magnificent white-knobbed Twins, the new Vibrolux is an amp to admire. The Alnico speakers should make it inviting to fans of the brown Vibroverb.

"Custom" Vibrolux Reverb, 95–. *The 2x10, two-6L6, tremolo- and reverb-equipped Vibrolux returned in '95, with a new look, going back to '63.*

VIBRASONIC

1959–1963
1972–1981
1995–Present

Vibrasonic
Vibrosonic Reverb
"Custom" Vibrasonic

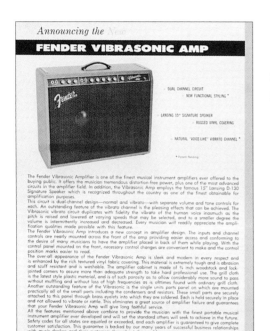

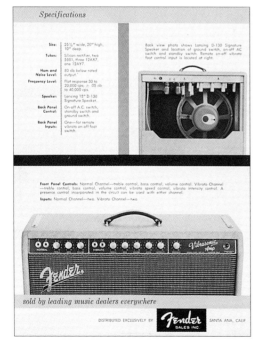

Brown Vibrasonic ('59), '59–'63. *Introduced front control panel, original-equipment J.B. Lansing speaker, Tolex, cylinder knobs, flat logo, and chassis straps. Wow! It also introduced a new style of tremolo. The grille cloth changed in '61 to maroon and again in '62 to wheat. Very expensive. Note prototype knobs and lack of separate back panel.*

The Vibrasonic amp is remembered today not so much for its tone or its association with famous musicians, but for introducing a number of features that would become standard appointments on Fender amps right up through today. The short-lived amp (model 5G13) was released in time for the National Association of Music Merchants (NAMM) trade show in the summer of '59, and it was new, new, new. New sounds came from a stock 15" J.B. Lansing D-130 speaker (a Fender first) and from the tremolo circuit, Fender's first in a large amp and considerably more elaborate than earlier designs. New looks included brown "textured vinyl" covering (instead of tweed), a brown plastic handle (instead of leather), and a brown control panel with brown knobs (instead of a chrome panel with black pointer knobs). The earliest versions used a slightly different brown knob than the style that became standard (see photo). New feel came from having the control panel moved from the top of the amp (with the lettering facing the rear) to the front of the amp (with the lettering facing the front). This feature kept players from having to get behind their amps or read upside down to change settings. Fender would go on to integrate this design into all its amps, and still continues the practice. (Note: In George Fullerton's book *Guitar Legends*, Leo Fender and the late, great Luther Perkins are pictured with a prototype Vibrasonic. The caption states, "The calendar on the wall says June, 1958." It's actually June 1959.)

The Vibrasonic, although equipped with only two 6L6 power tubes, developed a loud, clear tone, its big transformers and the efficient Lansing speaker making it louder than most amps with twin power tubes. Costing 20% more than the 2x12 four-6L6 tweed Twin amp, and weighing four pounds more, the Vibrasonic was heavy-duty. Two completely independent channels, Normal and Vibrato, each had separate Bass and Treble controls, a first for Fender. The volume controls, which were to the right of the tone controls

(Bass, Treble, Volume) on the early models, were independent of each other, another first for Fender. This was accomplished by using both sides of a 7025 for the preamp of the Normal channel, and another 7025 for the Vibrato channel. Half of a tube was used as the first stage, followed by the tone and volume circuitry, and then the second half of the tube. This isolated the controls from the following circuitry, keeping the volume of one channel from affecting the other, as on tweed amps. The *long-tailed* phase inverter, the Presence control on the negative-feedback loop, and the pair of fixed-biased 6L6GCs had become standard for Fender's larger amps, but the use of six silicon diodes in place of the rectifier tube was a first. Considering all the variations in rectifiers Fender had tried (one 5U4, two 5Y3s, two 5U4s, one 83 mercury-vapor tube, and the new GZ34), it suggests the designers were looking at this section as holding the large amps back. "Silicon rectifiers used in the circuits offer smoother power regulation and eliminate the heating problems encountered with glass tubes." A whopping 456 volts were now delivered to the power tubes—more than ever before.

The tremolo circuit was different than any earlier or later versions. Two twin triodes (7025s) worked just with the signal from the Vibrato channel preamp, allowing the Normal channel preamp to operate independently. Designed to get rid of the "ticking" sound when no musical signal was applied, the elaborate circuit included a filter network (crossover) that sent the lows to one half of a 7025 and the highs to the other half. 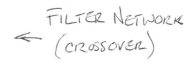 The oscillator section—half of another 7025—modulated the lows. It also fed the other half of that tube, which inverted the polarity of the oscillation. This inverted oscillation modulated the highs. All the information common to both lows and highs—e.g., ticking or hum from the oscillator—would be canceled when the two sections were combined (humbucking). The highs and lows pulsated back and forth against each other, a very complex but wild sound! Fender actually tried to pass it off as pitch-bending vibrato, but compare it to a Magnatone 260 or 280 and you'll hear that it's tremolo, albeit different from the earlier fixed-bias modulation or later grounded-out preamp signal.

An upgrade in the tremolo circuit requiring an extra 12AX7 (model 6G13-A) accompanied the change to controls in the sequence Volume, Treble, Bass. The grille cloth was changed to the new maroon color and the Tolex changed color slightly c. early '61. Another change in grille cloth to the wheat color seen on the small combos completed the transformation.

Starting with a high price that was certainly justifiable considering all the new features, Fender would never raise the price of the Vibrasonic, even when everything else in the amp line went up in '62. On the other hand, they never lowered it either. Fender introduced its "Professional" series in 1960, all nearly identical to the Vibrasonic in design and appointments. The unchanged prices between the relatively primitive tweed models and the modern brown Tolex amps made the new Pro, still at $284.50, a real bargain, especially when comparing it to the $479.50 Vibrasonic, which was basically built from the same schematic! The 15" J.B. Lansing was expensive, but the Pro's 15" Jensen certainly wasn't cheap. The transformers were different, but both put out the same high voltage (power transformer) and capably handled the amp's final signal (output transformer). Even the white Twin of '61, with two 12s and four 5881 power tubes, cost $50 less than the Vibrasonic. Fender Sales must have been unwilling to lower its price for fear of dealer resentment, instead choosing to leave the amp comparatively overpriced until it was effectively replaced by the 1x15 Vibroverb at $339.50—way more amp for way less money.

The Vibrasonic was Fender's top of the line for only a short period, being replaced there by the Showman at the end of 1960. It would stay as the number two amp, between the Showman and the Twin, until the release of the new blackface Twin Reverb, Super Reverb, etc. in the summer of '63, at which time it was retired.

In 1972 a new 100-watt silverface Vibrosonic (note the spelling) Reverb was released, similar to the old Vibrasonic but with more power. This was the same circuit design as the Twin, Quad, Super Six, and Showman Reverb amps. A heavy-duty Fender PS 15 speaker helped keep the sound clean, making it popular with pedal steel players, amongst others. This was Fender's first 1x15 combo amp since the black Pro was discontinued in '65. A JBL D-130F

Silverface Vibrosonic Reverb ('72), '72–'81. *When Fender decided to reinstitute the single-15" combo, they simply took a Dual Showman Reverb head, a JBL 15, and built a cabinet around them. The last of these came with blackface appointments. Popular with pedal steel players.*

was offered as a less expensive option to the PS 15 in '74 and soon was the stock speaker. In '79 the JBL was replaced with a 15" Electro-Voice, the last change to the Vibrosonic Reverb before it was discontinued in 1981.

A recent addition to the Fender line is the "Custom" Vibrasonic, "designed in the Amp Custom Shop but built on the regular line." This 100-watt tube amp with a 15" speaker is similar in conception to the Vibrosonic of the seventies, and features Reverb on both the "Steel" and "Guitar" channels. The blackface panel, black knobs, tilt-back legs, and silver grille cloth go back to a period when there was no Vibrasonic available, with only the rubber handle and flat logo going back to the original model. The tremolo (still being called vibrato; wake up, guys...) operates on both channels, as on the brown Vibroverbs (and unlike any Vibra- or Vibrosonic), but uses the photoresistor circuit instead of varying the bias to the power tubes. Having reverb on both channels is also unlike any of the amps that inspired this; however, the power amp and Steel channel preamp stages are almost exact replicas of the blackface Twin Reverb. An unbypassed cathode resistor on the first preamp tube brings the gain down a bit and reduces distortion, while a 50pF Sweet switch replacing the 120pF Bright switch adds a touch of sparkle to the extreme highs. These two minor changes, coupled with the 15", are a winning combination for steel players or jazzers seeking a clean, rich tone. A Fat switch, as seen on the Vibro King, etc., is a practical addition to the Guitar channel, which also uses slightly different values for the EQ section. So while this amp is certainly not a reissue of any one model, it is a well-thought-out combination of features, kind of a black Vibroverb with twice the power or a blackface Twin with a 15". A worthy successor to the Vibrasonic name.

"Custom" Vibrasonic, '95–. *Fender went straight for the new country market, having Steel and Guitar channels, both with reverb and tremolo. Otherwise, the basis for the rest of the circuit was the Reissue '65 Twin Reverb.*

CONCERT

1959–1965
1982–1985
1993–Present

Shortly following the first Vibrasonics, Fender released another short-lived amp with new looks, sounds, and feel. The Concert amp, equipped with two 6L6 power tubes and four 10" Jensen speakers, featured the new brown Tolex and all the *accoutrements* that went with it. Tremolo circuitry and a front-mounted control panel were a few of the other differences between the Concert and Fender's already existing 4x10 amp, the tweed-covered Bassman. Apparently the Concert was being worked on at the same time as the Vibrasonic, having a lower model number (12, vs. 13); but it would not be released until very late '59 (e.g., model 5G12, serial no. 00271, dated January '60).

Brown Concert ('60), '59–'63. *Joining the Vibrasonic in late '59, the Concert was basically a 4x10 Bassman for guitarists, with the Vibrasonic's updated circuitry. Originally fitted with tweed-era grille cloth, the amp changed to maroon in '61 and wheat in '62.*

The controls on the first version were mounted as follows (*L to R*): Bass, Treble, Volume, with Speed and Intensity on the Vibrato channel, and a master Presence control. Tubes were five 7025s and two 6L6GCs. The Concert (model 6G12) was exactly the same as the Vibrasonic except for the 2Ω output transformer and the necessary change in value of the negative-feedback resistor. Even with four 10s (24"x24"x10½" cabinet), the Concert cost 25% less than the Vibrasonic.

In late '60 the tremolo circuit added another tube (12AX7), and the controls were changed to Volume, Treble, Bass (model 6G12-A). The tweed-era grille cloth was replaced with the new maroon color in early '61, and a richer shade of brown Tolex followed a short time later. In '62 the grille changed again, to wheat-colored cloth; otherwise little was changed (black handle, smoother Tolex) until the blackface model in the summer of '63.

Black Tolex and knobs were a brand-new look, as seen in the *Down Beat* insert of September '63. The prototype model shown featured a flat logo and the solid gray prototype grille cloth. The '63–'64 full-line catalog showed the production model (AA763), with silver sparkle grille cloth and the new raised logo. Bright switches replaced the Presence control, and the phase inverter tube was changed to a 12AT7, as on all the blackface models of the time. The big change came in the tremolo circuit: the design from the brown models, with two 12AX7s and one 7025, was replaced with one 12AX7 and a photoresistor. This circuit dumped the preamp signal to *ground* instead of varying the grid bias. For barely 5% extra ($20) one could have bought a new Super Reverb (which should have been called the Concert Reverb); consequently, blackface Concerts were not big sellers. As with the Vibrasonic, Fender chose to maintain its price, and production was stopped before CBS took control. The Concert was last seen on the February '65 price list.

Blackface Concert ('64), '63–'65. *A new tremolo circuit, Bright switches, and black Tolex were the big changes for this short-lived version. The Concert had to compete with the new Super Reverb and was retired around the time of the CBS takeover.*

The name reappeared in '82 as part of the "II" series, which replaced the long-running line of combo amps equipped with reverb and tremolo. The new Concert was available as a 4x10, 2x10, or 1x12 combo, as well as a separate head. These new amps lacked the "unhip" tremolo circuit, opting for the "popular" high-gain, pull-knob EQ style of the time. Two parallel inputs fed a 7025 twin triode, with the two sides direct-coupled, and from here the signal path was split. A traditional Fender sound (minus tremolo) was available from the clean channel and its setup of Bright switch, Treble, and Bass, followed by the Volume control and half of another 7025. The Hot Rod channel sported a Mid control with a pull-knob Mid Boost to supplement its Bass and Treble controls. Like the clean channel, a Volume control and another stage of 7025 followed. This signal, however, fed another stage of 7025, with a level control pot in between

"II" series Concert ('82), '82–'87. *The 4x10 Concert returned in '82, with overdrive and channel switching replacing the tremolo. Most of these amps, however, had the 1x12 option.*

labeled Gain. The Volume and Gain controls allowed the last two stages of 7025 to be run wide open for maximum distortion. A Master Volume pot followed the final stage, and this high-gain signal joined the clean signal at the standard reverb section (12AT7 driver, 7025 recovery).

A buffered effects loop with level controls for the Send and Return sent the signal through three more stages of 7025, two more pots, and three more capacitors, not to mention the mysterious workings of the devices plugged into the loop...*or* all but one stage of 7025 could be bypassed by resisting the temptation to plug something into the loop. A standard 12AT7 *long-tailed* phase inverter and the return of the Presence control on the negative-feedback loop completed the signal path to the 6L6s.

Things had changed very quickly for the Fender line, which had resisted change for almost 20 years. The Concert was the only amp with two 6L6s that Fender made between the time of its new release and the introduction of the Super 60 in 1988, making it a very important amp in the history of Fender. Construction had ceased with the sale of the company by CBS, but existing inventory lasted into 1987.

Concert ('93), '93–. *Returning in the popular 1x12 setup, the Concert offers a mix of classic blackface and modern design.*

A 1x12 two-6L6 Concert joined a new 4x10 Super in replacing the "Red-Knob" series Super 60 and the other "Super Series" amps. Released in 1993, these "Professional Tube Series" amps, joined by the updated Twin amp in '95, represent the progressive side of Fender's designers, offering a high-gain tube amp to complement the Reissues and Custom Amp Shop models. Two selectable channels—Normal and Drive—with separate tone controls, Reverb, Effects loop, and a Line Out jack, make the 60-watt 1x12 Concert appealing to players who need a modern amp but also appreciate the necessity of a good "clean" sound, something many high-gain amps dreadfully lack. A foot switch activates three different settings: the Normal channel with Treble, Bass, and Mid controls, or one of *two* levels of Drive channel, which share a common set of Treble, Bass, and Mid controls. A master Volume for the Drive channel works with two nearly identical Gain controls, allowing for one to be set appropriately louder than the other. The second Gain circuit includes the additional equivalent of a Fat switch. While there is minimal difference between the tonality of Gain 1 and 2, the Drive channel does use one more preamp stage than the Normal (as well as having the master volume).

An Effects loop common to both channels has a Mix control on the front panel. This control keeps the majority of the original signal inside the amp and out of the tone-devouring effects boxes. The effects can be set totally "wet" (no original signal) and then blended in, via the Mix control, with the pure, unadulterated, all-tube "dry" signal. This parallel signal flow is more like the way effects are added on in the studio, as opposed to the series flow traditionally found in effects loops, which makes the entire signal pass through the external solid-state effects devices. Three levels, from −24 to +2 dB, allow the use of any external unit. In situations where parallel operation is inappropriate (e.g., tremolo or intense flanging), turning the control fully clockwise (to 10) gives a traditional series effects loop. Also common to both channels is the spring reverb, which uses a 12AT7 driver and a 12AX7 for the recovery. A 12AX7 long-tailed phase inverter connected to the negative-feedback loop feeds the power tubes, and a solid-state rectifier powers the DC operations. The basics of these amps are classic Fender with just enough practical extras; simply a great design.

SHOWMAN
1960–1981
1987–1993

Showman
Dual Showman
Dual Showman Reverb
Dual Showman SR

Electronically, the Showman amp released in December 1960 offered nothing new, since it combined the operations of the brown Tolex "Professional" series with the four-5881 power of late-fifties Twins. It was and is a *very* important amp, however. Besides introducing the tasteful attire of white Tolex, maroon grille, and white knobs, the Showman was Fender's first production "Piggyback" model. Although this development is often attributed to making a heavy amp easier to carry by splitting it into two separate pieces, examination of the early Showman bottoms clearly shows a more scientific purpose. The powerful "head" had been perfected for the Twin, but finding speakers to handle the massive power output proved difficult, especially in an open-back cabinet.

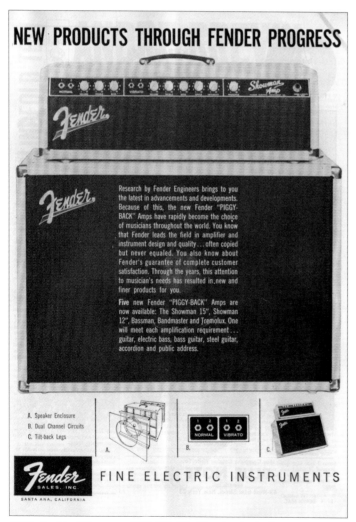

White-knobbed Showman ('61), '60–'63. *The piggyback Showman was the direct result of Fender's trying to contain its four-5881 tweed Twin. The new ported, JBL-equipped cabinet was available in two sizes: 1x12 and 1x15. Changed to wheat grille in '62, just after the release of the 2x15 "Double Showman."*

The new separate enclosure featured an elaborate porting system, with the stock Lansing speaker mounted in a metal *tone ring* that was attached to a separate inner baffle board. This, in turn, was mounted on the regular baffle board, which was flush with the front of the cabinet. In mounting the inner board to the outer, air space was left on the sides for low-frequency waves coming off the back of the speaker to escape, hopefully in phase with the waves coming off the front of the speaker. This "bass reflex" method improved low-frequency response greatly, allowing the player to back off the amplifier's Bass control, freeing up more power for the middle and high frequencies. The result was loud, full-range, distortion-free sound. The cabinet was equipped with "tilt-back" legs, allowing it to be placed on the floor—again improving bass efficiency—but leaning back

110

so the speaker was aimed at the musician's head—POW! For years nothing could touch this amp, and to this day it is a favorite for a number of artists.

On the other hand, the Showman was a roadie's nightmare. While the Twin, the 4x10 Concert, the Lansing-equipped Vibrasonic, and the 4x10 Bassman all had shipping weights between 50 and 60 pounds, the Showman 15 left the factory at 110 pounds! Also available was the 100-pound Showman 12.

Both cabinets were fitted with a pair of T-nuts on the top that accepted two bolts with large knurled tops, similar to Tele knobs. The head had two brackets on the bottom that slid out, allowing it to be bolted securely to the bottom and have both pieces safely tipped back, defying gravity. Or the user could place the head away from the bottom, either for convenience or to spare it the vibrations of being mounted on a cabinet putting out 120 decibels. The piggyback idea went back to a two-piece 4x10 Bassman special-ordered a few years earlier (see page 185), and a few prototype piggybacks were covered in brown Tolex, which can be traced to the limbo era of the Twin (see Twin). Dick Dale was reportedly a recipient of a brown Showman in 1960, and the catalogs from that year picture a brown piggyback being used by the Champs of "Tequila" (and "Subway," "20,000 Leagues," etc.) fame (see page 30).

The heads of the '61 Twin and Showman were identical, as they would be for most of the next 20 years. (The exception was the period from mid '63 to mid '68, when the Showman lacked the addition of reverb.) Inside the amp (model 6G14) was the new tremolo circuit with three tubes, as seen on the second-version Vibrasonic, etc. The power supply included the same power transformer, silicon rectifiers, and filter caps as the Vibrasonic, plus the same phase inverter circuit (7025). The Showman, however, used four 6L6GC power tubes instead of two. The input sections were slightly different, but only in the values of the components; the tubes (two 7025s) and controls (Volume, Treble, Bass, Speed, Intensity, and Presence) were the same. Ground, On/Off, and Standby switches, a fuse holder (3-amp), speaker and extension speaker jacks, and an RCA jack for the Vibrato Pedal were fitted on the back panel. A second version (model 6G14-A) replaced the 6L6s with 5881s and two of the 7025s in the tremolo circuit with 12AX7s.

DOUBLE SHOWMAN 15" AMPLIFIER
Available by custom order only is the new Double Showman 15" Amplifier. The Double Showman contains the same amplifier section that is employed in the regular Showman, however it is equipped with two 15" J. B. Lansing Speakers.

This Amplifier was originally produced for experimental purposes but due to the response, tone and power, demand has been created. We believe the Double Showman to be the ultimate in amplifier performance and as such is guaranteed for any musical instrument or microphone. Amplifier shipping weight 125 pounds. Size: Same as Showman 15". Price: $800.00.

Double Showman ('62). Fender Facts #1 announced the release of what would soon be referred to as the Dual Showman, a Showman head with a 2x15 cabinet loaded with JBLs. (Courtesy Hardtke Archives)

As if this amp wasn't enough, the addition of an "improved" model was announced in the debut issue of *Fender Facts* (December 1962). The "Double Showman 15" amplifier used the same head as the regular Showman 15, and the bottom was the same size. But instead of the elaborate porting system, the cabinet was divided into two separate sections, each housing a 15" Lansing mounted directly on the front baffle board. While the ported 1x15 cabinet used progressive hi-fi design and theory to efficiently match the bottom to the head, the sealed 2x15 cabinet simply overwhelmed the capabilities of the head. The earliest version(s) still had maroon grille cloth, but these are extremely rare in original condition (*caveat emptor*). By the end of '62 all the white Tolex amps changed to wheat grille cloth. The brown handle was replaced with a new black plastic one in the spring of '63, just before the white-knobbed Showman was last seen.

The *Down Beat* insert of September '63 showed the Dual Showman with black-skirted knobs and the new circuit (AA763) that replaced the Presence control with Bright switches on each channel, specified 6L6s for the power tubes, and replaced the 7025 phase inverter tube with a 12AT7. A new Middle control was added, and the plate voltages on the preamp tubes were increased 50%, making these amps much cleaner, with more headroom. Another big change was in the tremolo circuit, with one 12AX7 and a photoresistor replacing the three-tube version. This tremolo design connected the preamp signal to the Intensity pot and from there to the photoresistor. This device varied its resistance in sync with the tube oscillator, allowing the preamp signal a path to *ground*. A 10kΩ *bias adjust* pot was added to the fixed-bias supply for the power tubes. Like the Bandmaster and the Tremolux, the last of the white Showmans were more like the black Tolex blackface Showmans than the earlier versions.

A raised logo, smooth white Tolex, and gold sparkle grille cloth were the last changes (by late '63) before the Showman joined the combos with black Tolex and silver grille in mid '64. For a time during the Fender Musical Instruments years, the Dual Showman actually had the word "DUAL" in large block letters added above the script "Showman Amp" logo.

So was there a difference between Showman and Dual Showman heads? Yes. On the models with Bright switches, the single-speaker version used output transformer #125A29A and a 47Ω resistor where the phase inverter and negative-feedback loop tie to ground; the two-speaker version used output transformer #125A30A and a 100Ω resistor. On the models from late '62 to mid '63, the Dual Showman *probably* had a different output transformer than the #125A4A used on the single-speaker version. (Anybody know the number?)

The Showman 12 was dropped in the summer of '66, coinciding with the release of the Solid-State Dual Showman. The last of the blackface Dual Showmans had an unnecessarily huge 2x15 cabinet (45½"x30"x11½"), about 50% larger than the earlier model. In the summer of '68 the Dual Showman Reverb head was released (model AA768), with silverface panel and aluminum trim, returning to the original idea of a Twin head (this time Twin Reverb) in a separate cabinet. Why Fender maintained a separate reverbless chassis for the Showman from '63 to '68, especially since the Showman was the top of their line, is a point of contention (see Bandmaster Reverb). The regular Dual Showman was quickly phased out, and by the end of the '68 so was the Showman 15. The Showman and Dual Showman ($40 extra) heads were still available separately through the beginning of 1970.

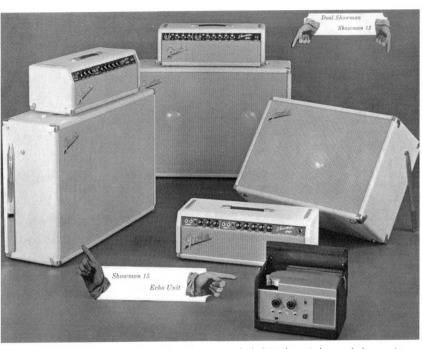

Blackface Showman ('63), '63–'67. New circuitry included Bright switches and photoresistor tremolo. The grille cloth would change to gold, and in '64 the Tolex changed to black. The Showman 12 was discontinued in '66, while the last Dual Showmans came with a giant speaker bottom.

The first silverface Showmans were almost identical to their blackface predecessors, but around mid '68 and the release of the Showman Reverb, some big changes occurred, particularly in the bias section of the power amp. Tying the grids to ground with capacitors, lifting the cathodes above ground, and changing the 10kΩ bias adjust pot to a balance pot were a few of the changes that gave these amps (models AC568 and AA768) a bad reputation. The problem was somewhat remedied the following year, but the balance pot

FENDER AMPS: THE FIRST FIFTY YEARS

Silverface Showmans, '68–'81. The Dual Showman and Showman 15 did not last long, being overshadowed by the new Dual Showman Reverb, released in '68. This amp was simply a Twin Reverb in piggyback clothing. Retired in '81.

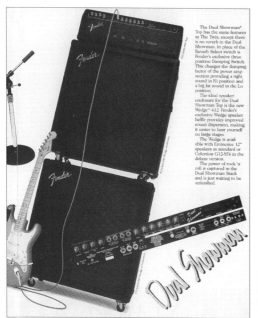

stayed. The aluminum trim was dropped and a new three-position ground switch accompanied the grounded AC plug c. 1970. A Master Volume circuit and a Hum Balance control were added c. '72, and removable casters became standard on the cabinet. By '76 the logos had lost their tails, and a pull-knob Boost pot became part of the Master Volume circuit. The silverface Dual Showman Reverb, also referred to as the TFL5000D on early versions, was Fender's top guitar amp until the release of the 300 PS in 1975. The 1980 line showed the Fender 140 replacing the 300 PS as Fender's big progressive amp and the traditional Dual Showman Reverb returning to its previous position at the head of the price lists (even though the 140 full stack was more expensive). In 1981 the last of the original Showmans, as well as the Bandmaster and the 140, were retired, leaving the new 75 as the only Fender guitar head. By '83 the Showman name was revived as a solid-state amp. These were offered during the transition from CBS, and leftover units were sold into 1987.

A new tube Showman—a reverb-less piggyback counterpart to the new "The Twin"—marked a return to the old tradition in 1987. One hundred watts of tube power, plus channel switching, effects loop, high-gain preamp, and a three-position damping control, made the new model worthy of the name. Cabinets containing four 12" speakers replaced the old 2x15 JBL setups that Fender had stood by all through the seventies and early eighties when 4x12 bottoms had become the standard. These Showmans were part of the new small line of tube amps from Fender Musical Instruments that became known as the "Red-Knob" series.

The Dual Showman SR with Reverb replaced the previous model in 1990 and was upgraded in '92 with a 25-watt switch and parallel channel mode, allowing all the controls to be operational at once. By '94 and the release of the Custom Amp Shop Tonemaster, the Dual Showman was again laid to rest.

Dual Showman ('87), '87–'93. A modernized Showman appeared in '87, with distortion, 4x12 bottoms, and no reverb. Red knobs were standard, and custom coverings were optional. Reverb was added in '90, but the amp was discontinued in '93.

VIBROVERB
1963-1964
1990-Present

One of the most influential Fender amplifiers of the sixties was the Vibroverb, being the first model from the company to feature built-in reverb. Officially announced in *Fender Facts* #2 (February '63) and listed in the February '63 price list, the brown-Tolex-covered classic, equipped with two 10" speakers and a pair of 6L6GC power tubes, was not around long enough to have much impact on the market; but Fender knew they had a winner. Based on the 2x10 Super of the time, the Vibroverb (model 6G16) shared similar input sections (7025 Normal, 7025 Bright), phase inverter (7025), power amp section (two 6L6GCs), and rectifier (GZ34). Controls for the Normal channel were Volume, Treble, and Bass, and for the Bright channel Volume, Treble, Bass, and Reverb. To make room for the Reverb, Fender reverted to a tweed-era tremolo circuit using a single-tube oscillator (12AX7), varying the fixed-bias supply for the power tubes. This method imposed tremolo on both channels or neither, but required two fewer tubes than the Super's. These two tubes would be used for the reverb driver (12AX7) and recovery (7025) sections. The speakers on production models were manufactured by Oxford and had brown molded plastic covers. They appear to all be from the same batch, as numerous examples (e.g., serial no. 00113, dated May '63) have date codes of 465311, implying a spring '63 release.

Brown Vibroverb, '63. *Fender's answer to the myriad of reverb-equipped amps already on the market by early '63. This amp didn't make much of a splash until collectors picked up on its unique combination of features.*

By the end of the summer, the Twin Reverb (2x12, four 6L6s), Super Reverb (4x10, two 6L6s), and Deluxe Reverb (1x12, two 6V6s), all covered in the new black Tolex and sporting silver grille cloth, offered the public a variety of self-contained amps with reverb. This marked the end of the brown Vibroverb. Rumors exist of prototypes from late '62 covered in white Tolex, but the short production run was consistent, with the exception of the earliest model(s) having a leather handle instead of the usual black plastic.

Blackface Vibroverb ('64), '63–'64. *A blackface Pro with reverb. A 1x15 cabinet and black Tolex were two of the big changes for the Vibroverb. Phased out in less than a year.*

The *Down Beat* insert of September '63 did not include a Vibroverb, implying a temporary halt in production; but the '63–'64 full-line catalog pictured a two-6L6 1x15 version. Covered in black Tolex with silver grille, it replaced the original, if only in name. This version (model AA763), also very short-lived, was different from its brown predecessor in more ways than the new cosmetics and different speaker size. The tweed-era tremolo circuit was replaced with the new photoresistor version (12AX7), which affected only the Vibrato channel. Also new were Bright switches, common to the larger blackface amps; otherwise the controls were unchanged. This amp should have been called the Pro Reverb, since it was essentially a blackface 1x15 Pro with reverb; witness the nearly identical inputs sections (two 7025s), tremolo circuit (12AX7), phase inverter (12AX7), power amp section (two 6L6s), rectifier (GZ34), fixed-bias supply, and transformers. Instead, Fender kept the Vibroverb name through the fall of '64, when it was retired. The 1x15 combo, popular for years with "professionals," was phased out in early '65 when the 1x15 Pro was replaced with the 2x12 Pro Reverb.

The two Vibroverbs are both sought after today, partially for their scarcity (and hence collectibility), but also because they offer combinations of features not available on any other combo. The 2x10 model is unique for its historical standing, cosmetics, and tube tremolo. The blackface model was the only 1x15 combo with reverb until the 100-watt Vibrosonic was released in 1972. Both Vibroverbs are great amps, but for totally different reasons.

The importance of introducing reverb into the Fender amp line cannot be discounted, although it is more important to the history of Fender amplifiers than to the history of amplification in general, as the Vibroverb was actually late in joining the party. Danelectro, Ampeg, Epiphone, Supro, Harmony, Kay, Rickenbacker, National, Guild, Magnatone, and Gibson all offered their own reverb combos before the Vibroverb. But which would you take to a late-night gig?

As demand for original Vibroverbs increased in the eighties, Fender responded with a "Reissue" version in 1990. A number of differences—e.g., printed circuit board, ¼" reverb jacks, etc.—keep any confusion from arising as to when a version was built, but the attention to cosmetic details and, most importantly, the *tone* have made the reissues a big success for the new company. The circuit is close to identical except for the tremolo section, which uses components of different values but in approach is the same as the original. A solid-state rectifier is standard in place of the original GZ34. If nothing else, the Reissue Vibroverb should be hailed for reintroducing tremolo to the line following an extended period in the eighties when it was not available on *any* model.

Reissue '63 Vibroverb, '90–. *To accommodate the demand for original '63 Vibroverbs, Fender chose this model, along with the Bassman, to inaugurate their Reissue series.*

BRONCO
1967–1975

When is a Vibro Champ not a Vibro Champ? When it says "Bronco" in red letters! Fender wanted a new amp in the student range to go with their new Bronco guitar, announced in 1967. According to the July Price List and *Fender Facts* #14 (July '67) the set was available (no prices yet), but in reality the Fender factory was having trouble keeping up with the demand for its more popular models. The *Fender Facts* showed a blackface amp with the name covered by the Bronco guitar, implying the prototype Bronco amp was not ready. *Fender Facts* #15 (November '67) again showed an amp with the Bronco guitar and noted that the guitar would be available in November and the amplifier in December. This prototype amp had *four* black plastic "F" knobs with metal skirts and a silver control panel. The logo was missing its tail and there was no aluminum trim around the grille cloth, unlike the solid-state amps of 1966 and the tube amps of '68. To further confuse the issue, the 1968 full-line catalog again showed the Bronco guitar covering the name of a blackface amp. This catalog, however, should say '67–'68, as it was prepared in the summer of '67.

Prototype No. 1 ('67). *First view of the alleged Bronco (spelled "C-h-a-m-p").*

Prototype No. 2 ('67). *From the '68 catalog. Another "cleverly disguised" C-h-a-m-p.*

Prototype No. 3 ('67). *Note only four knobs and early use of silverface on a tube amp.*

The 1969 full-line catalog ('68–'69) finally showed the production-model Bronco amp. It had the standard features of the era: silverface control panel, "blue" grille cloth with aluminum trim, and *five* black-skirted knobs. The controls were identical to those of a Vibro Champ—i.e., Volume, Treble, Bass, Speed, and Intensity; schematics and manuals listed the Vibro Champ and the Bronco as the same amp. The lettering was in red instead of blue—very sharp! The logos on all the small amps and some of the larger ones that year were tail-less, throwing a monkey wrench into the logical progression of features.

By 1970 the tail was back on *all* the logos, and the aluminum trim around the grille cloth was gone. This would be the standard Bronco through early 1975, when the amp was discontinued. The red Bronco guitar was also discontinued in early '75. (Perhaps they ran out of red paint?) Bronco guitars were available from 1975 through 1980 in black or white.

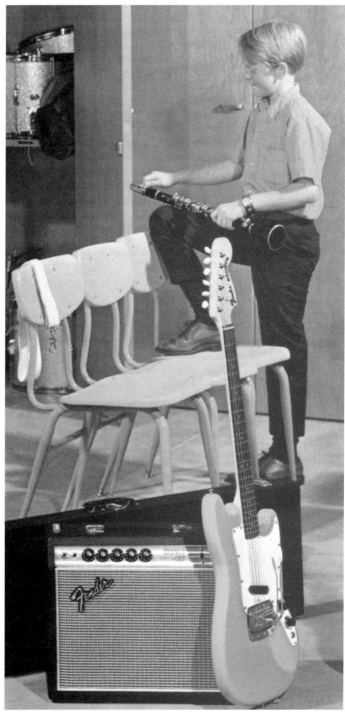

Early production-model silverface with aluminum trim ('68). *It lost the trim c. '70 and was discontinued in '75.*

BANTAM BASS
1969–1971

In an early example of East-meets-West technology, Fender offered the short-lived Bantam Bass amp. The box and amp were standard fare for the time, but turn that amp around and you'll see one of the strangest and least logical looking speakers in the history of electronics, courtesy of Yamaha, who were brazen enough to put their name on it! The cone was made of white Styrofoam, and—it gets weirder—the shape defies any sane description. What were they thinking in Yamaha R&D, and how was CBS talked into buying these? I can hear the salesmen now: "If you ever blow one of these [they all blew], replacements will be readily available." Needless to say, working examples of the Bantam are quite rare today.

The amp had a Bass Instrument channel with Volume, Treble, and Bass controls, as well as a Deep switch. The Normal channel also had a Middle control and substituted a Bright switch for the Deep switch. The electronics were standard Fender circuits: two 7025s for the preamps, a 12AT7 phase inverter, two 6L6GCs for power, and a 5U4 rectifier. Replaced by the Bassman 10.

From the front, the Bantam Bass could pass for any number of Fender amps of the period...

...but turn that amp around....

MUSICMASTER BASS
1970–1982

Following the disastrous release of the solid-state amps, the consensus of a degradation of the tube amps, and only two new tube amps in five years (the uninspired Bronco and the out-of-this-world Bantam Bass), it appeared that Fender under CBS was having some difficulties. But with the release of the Musicmaster Bass amp, they offered something inspiring: a solid little amp designed especially for bass.

Both sides of a 12AX7 were used as the preamp section, and a pair of 6V6GTAs were used for power. The solid-state rectifier was nothing unusual, but the phase inverter was. Fender used a transformer with a center tap on the secondary, which gives two theoretically identical but out-of-phase signals…the perfect phase inverter! This is actually one of the earliest methods for this function. So why didn't Fender use it on all their amps? Probably for the same reason the other tubes were capacitor-coupled and not transformer-coupled: cost. While the tube phase inverter circuit is certainly more elaborate and labor-intensive than the installation of a simple transformer, a high-quality balanced transformer is still the more expensive method. It's interesting that Fender chose this amp for their experiment.

Volume and Tone were the only controls; the knobs were black plastic with metal skirts. Later seventies versions have a tail-less nameplate, and the blackface control panel classed up the amp for its final year in '82.

A need was filled by this 12-watt amp with its heavy-duty bass speaker, and one could do a whole lot worse when looking for a small bass amp for practice or recording.

Early model, silverface with metal-skirted knobs ('70). *Last models ('81–'82) were blackface. Replaced by the Bassman 20.*

400 PS BASS
1970–1975

Through 1969, the 50-watt Bassman was Fender's idea of a bass amp. Having the bass player use the guitarist's 100-watt Showman and trading him the 50-watt Bassman for guitar was a practical but unofficial solution to the problem of not enough power for loud, clear bass. In '69 Fender released the 100-watt Super Bassman, but Sunn, Marshall, Kustom, Plush, Acoustic, Standel, Sunn, et al. were all offering bass amps with at least twice that power. In 1970 Fender's answer, the 400 PS Bass, was released, comprising three separate 145-watt power amp sections: with one speaker, 145 watts were available; with two speakers, 290 watts; with three, 435 watts! That's RMS! Fender claimed 980 watts peak power available! The bottoms were heavier-duty than regular Fender bass bottoms, using an 18" bass speaker with an efficient folded-horn design.

Fender now offered what the public had asked for, but they had, perhaps, over-designed the amp. Two 7025s for the preamps, a 12AT7 and a 7025 in the reverb circuit, and a 12AX7 for the tremolo were all standard Fender design, but the power section of this behemoth was unlike anything before or after it. An *extra*-huge power transformer produced a heart-stopping 700 volts for the plates of the six 6550s (a tube previously ignored by the design team). This configuration had been used by Ampeg for their SVT amp. The primary of the output transformer was center-tapped for push-pull Class AB operation (three 6550s per side), and the secondary had three separate windings, splitting the 435 watts into three manageable

The first large amp by CBS, and the largest ever made by Fender. First style ('70–'72) used regular grille cloth.

outputs. In theory, these could be jumpered to form one monster output, but the individual secondary windings probably would not take the whole amount. Besides protecting itself, the three-section transformer protected the speakers of the world from being devoured by this beast. Ampeg later used a similar feature on their SVT, making the player use both the speaker and extension speaker jacks to receive the full output.

Here's how Fender explained things: "A single speaker enclosure large enough to utilize 435 watts, without damage to the speaker would be too large to be portable. For this reason, we have divided the available power into three separate speaker output jacks. This feature enables the user to alternate jacks for triple output tube life." This last sentence is an outright falsehood, and whoever was responsible for it owes an apology to anyone who has diligently alternated jacks. If there is anyone who has been doing this since 1970, you can stop now! The six power tubes were split, with the outputs of three paralleled to one leg of the primary and the second three paralleled to the other leg. The tubes saw the reflected impedance of the speaker, but it made no difference to them where you plugged in. The ultimate motive was written between the last lines of the amp's introduction: "Another speaker enclosure will double the effective wattage to 290. Using a third speaker enclosure, the full 435 watts will be realized." Let's see, *another* $500, and *another* $500....

In most Fender amps the phase inverter is a twin-triode tube, also acting as a driver. On the 400 PS, a transformer was used as the phase inverter, and this was driven by a 6L6GC; that's right, a 6L6 power tube on the primary side! The 6L6 was fed by a 12AX7 that summed the two preamp sections. Features included two

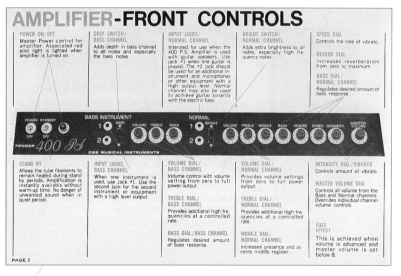

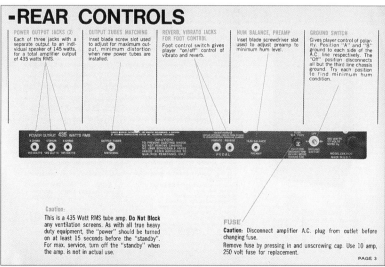

Master Volume was new to Fender tube amps. Though fitted with controls necessary for guitar, the amp was promoted mainly for use with the electric bass.

channels with a Master Volume, which they claimed would give a Fuzz Effect. The Bass Instrument channel had two inputs, a Deep switch, and Volume, Treble, and Bass controls. The Normal channel also had two inputs, followed by a Bright switch and Volume, Treble, Middle, Bass, Reverb, Speed, and Intensity controls. There was no mention on the manual cover (which says 400 PS Bass Amplifier) or in the guitar amp section of the '70 catalog about using this for anything but bass, but a lot went into designing this amp, and it was probably *not* so bass players could have Tremolo and Reverb at their fingertips, although this could be interesting. "Intended for use when the 400 PS Amplifier is used with guitar speakers" was all that was said. Using a guitar through the folded horn would certainly not be a treat! Two years later the catalog didn't even mention using it for guitar.

Fender offered the 400 PS through 1975, with the early models having "blue" grille cloth, and later versions (by '72) having black acoustic foam with white trim. The amp package included at no extra charge a four-wheeled cart to set the head on. Either as a marketing ploy exploiting the amp's enormous low-end output, or at the insistence of the "responsible" engineering team, Fender made the claim "For MAXIMUM Service Life. The amplifier should be operated in the stand, definitely NOT ON TOP of the speaker." Unfortunately, the stand looked more like it was borrowed from a caterer than purchased from a music store. This may be a logical design, but it lacked the almighty cool factor that people had gotten used to, seeing Entwistle, Redding, Wyman, et al. with their stacks.

These would be good amps for bass players to look for today, to go with the bottoms currently available that are both small and capable of handling high power. To build a tube amp with those transformers today would be an expensive proposition.

Restyled grille "cloth" was actually black acoustic foam, trimmed in white. Replaced by the 300 PS in '75.

SUPER SIX REVERB
1972-1979

The Super Six Reverb was a variation on the Twin Reverb using a cabinet with six 10" speakers instead of the usual two 12s. The cabinet was so tall they didn't bother putting a handle on the top, but included two on one side and recessed sockets for the removable casters on the other.

Schematics and manuals for the mid-seventies Twin Reverb, Showman Reverb, Vibrosonic Reverb, Super Six Reverb, and Quad Reverb are interchangeable, the only difference being the output transformers. Fender had always paralleled 8Ω speakers together, and used the appropriate output transformer, even on the 2Ω 4x10 Bassmans, Super Reverbs, etc. For six speakers, they had to do a series-parallel connection, which yielded approximately a 5Ω load. Fender called it 4Ω and used the Twin transformer.

Having a wall of speakers connected to the dynamic power of a 100-watt tube head makes for a very *big* rhythm sound, and the Super Six does it like no other combo. A thrill to be experienced! Introduced in February of '72 and available until the first part of '79.

One of five amps using the chassis of the Twin Reverb, the Super Six Reverb was virtually identical, but with six 10s instead of two 12s—in a series-parallel connection, unique for Fender. Phased out in '79.

QUAD REVERB
1972-1979

Like the Super Six and Vibrosonic, the Quad Reverb was a silverface Twin Reverb in a different cabinet, this time with four 12" speakers. And again, to hear the Classic Fender Sound, but moving that much air....

The Quad Reverb used something besides 8Ω speakers, very un-Fender-like. Four 16Ω speakers were paralleled for a 4Ω load, the same as the Twin Reverb, Dual Showman Reverb, and Super Six Reverb. Like the Super Six Reverb, the Quad Reverb was introduced in February of '72 and available until the first part of '79.

A 4x12 combo variation on the Twin Reverb theme.

SUPER TWIN
1975–1980

Super Twin
Super Twin Reverb

By the mid seventies, the Twin Reverb was still popular with working musicians, but it was time for Fender to release an improved model, to offer a step up. The Super Twin used two things that had always worked for Fender in the past: more power (180 watts compared to 100) and more tone controls (10 instead of five). The late-fifties Twin, with four 5881 power tubes and separate Bass, Middle, Treble, and Presence controls with Bright or Normal inputs, had long been considered the pinnacle of amplifier evolution. The addition of reverb and tremolo in the early sixties was icing on the cake. Were musicians asking for more from a Twin Reverb, or was Fender trying to entice players into a new, "improved" amp?

With the Super Twin, an elaborate equalization system was designed that included Bass, Middle, Treble, and Presence controls, a Bright switch, and five bands of fixed-frequency EQ (±5 [dB?] at 100, 235, 485, 1250, and 2300 cycles per second). The Bass, Middle, and Treble controls and the Bright switch were traditional Fender circuits, but the Presence and the five-band EQ were "active," very nontraditional for the company. The new Presence control was totally unlike the earlier circuit, which had changed the negative feedback in the high frequencies but was not really EQ. The Presence on the Super Twin was a boost at 3.9 kHz—definitely equalization. The five fixed frequencies differed from the Presence in that they were able to both boost and cut. Five *series resonant circuits* were constructed using an inductor, a capacitor, and a special potentiometer with a center tap for each band. One leg of each pot was connected directly to the signal path. This created a low-impedance path to ground for the frequencies surrounding (and including) the center frequency of that particular band (passive cut). The other leg of each pot was connected in parallel to the cathode bias resistor of a 12AU7. By lowering the resistance at the center frequency, the current and the gain of the tube

The Super Twin ('75–'76) was the first in a long run of high-powered amps with a leaning towards overdrive. Replaced by the Super Twin Reverb in early '77. The EQ section on these amps was more elaborate than on any other Fender, before or since.

increased around that frequency (active boost). This clever circuit used none of the usual op amps normally associated with active EQ and was a fairly expensive addition (quality inductors don't come cheap!). A foot switch could punch the EQ in or out, allowing the player two distinct tones and giving the Super Twin quasi-channel-switching abilities.

What players really wanted to switch, however, was distortion characteristics. The Super Twin was equipped with a built-in distortion control to do just that, in conjunction with an output level control on the amp that was activated by the distortion foot switch. A sliding on/off switch on the pedal allowed the output control to be activated or bypassed through the distortion foot switch at any level of distortion, including none. Despite the Master Volume and Fender's claim of "any desired degree of distortion...at any volume level," the distortion and sustain of this amp were not as intense as those of, in particular, the new MESA/Boogie amps of the time. And the weight was enough to do damage to all but the burliest, leaving Fender with another "ultimate" amp that was not received in the manner the company had planned. Although reverb and tremolo were considered somewhat passé at the time, their exclusion left many players with a feeling the amp was incomplete.

A solid-state rectifier powered the amp, and four stages of 7025 (i.e., two complete twin triodes) were used for the preamp section. The phase inverter (12AX7) was the standard *long-tailed* version, but instead of supplying the power tubes, it was followed by a 12AT7 in a cathode follower circuit. The low impedance of this circuit was a better match for the six 6L6 power tubes than the 12AT7 phase inverter Fender normally used.

Even more un-Fender-like was the choice of tubes for a revised model, released in early '77, that was equipped with reverb.

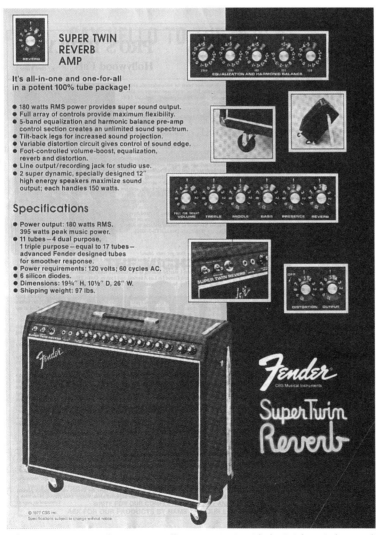

The Super Twin Reverb was essentially a Super Twin with the Bright switch moved to make way for a Reverb control. It incorporated two unusual tubes: a 6C10 triple triode and a 6CX8 pentode/triode.

The Super Twin Reverb amp used rectifier, power tubes, phase inverter, and preamp stages similar to the Super Twin, with a change to the last two preamp stages. The second 7025 was replaced with two sections of a 6C10 triple triode! The third section was used for the reverb recovery stage. Another oddball tube was used for the reverb driver: a 6CX8 pentode. What's even stranger was this tube had a second half—a triode—that Fender used in place of the 12AU7 in the equalizer section. Someone had their thinking cap on backwards when they worked out this design. The 6C10 was a recently designed tube, and neither it nor the 6CX8 is what you'd call common. If I had one of these amps, I'd be stockpiling these two tubes while I could. The Super Twin Reverb was phased out in 1980 with the introduction of the high-gain 30, 75, and 140 amps.

300 PS
1975–1980

A short time after retiring the 400 PS Bass in 1975, Fender released a more mainstream powerhouse, the 300 PS. Designed for bass guitar, the 300 PS was available with a special bass reflex cabinet with four 12" speakers angled from the corners inward slightly, as if aimed to fire at a common point three feet in front of the cabinet. For use with guitar, a smaller 4x12 cabinet with a flat baffle board was available. (Had Fender offered a special guitar cabinet for the 400 PS, it might have fared better.) The 300 PS head had a 300-watt output capability (660 watts Peak Music Power) and featured the controls first seen on the Super Twin: Volume, Treble, Middle, Bass, and Presence, with five bands of active EQ. The EQ knobs were numbered from –5 to +5 (dB?), and the entire section was turned on and off with a foot switch. Distortion and Output (Master Volume) were also foot-switchable. Gone from the 400 PS were Reverb and Tremolo. *Four 6550s were fed by a transformer phase inverter, this time driven by a 6V6GTA.* In conception, the output section of the 300 was very similar to the 400—i.e., 700 volts on the plates of the 6550s, solid-state rectifier, etc.

This amp was offered through the release of the solid-state B-300 in 1980. Another head for the smart bass player to consider. These large tube amps eat up similarly powered solid-state heads. *No comparison.*

Whereas the 400 PS was aimed almost entirely at bass players, the 300—which lacked the tremolo and reverb of the 400—was aimed at both bass and guitar players. For use with guitar, the amp was sold with a 4x12 cabinet. For bass, again a 4x12 cabinet, but one with speakers in each of the four corners, angled in towards the middle.

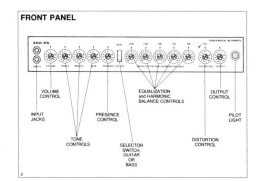

The 300 used the same controls as the Super Twin.

STUDIO BASS
1977–1980

This short-lived combo amp featured the elaborate EQ of the 300 PS and Super Twin, but joined the line near the end of their run and unfortunately was discontinued with them in 1980. Featuring 200 watts of tube power and a heavy-duty 15" Electro-Voice speaker, this little-known amp should have been successful for Fender, but was pulled before it could establish itself.

Six 6L6GCs were supplied high voltage (500V) from a solid-state rectifier, and a fixed bias ran through an "Output Tubes Matching" control on the back panel. Two 7025s were used in the preamp section, with a 12AU7 driving the equalizer section. The phase inverter circuit was more elaborate than usual, due to the large number of power tubes. A regular *long-tailed* inverter (7025) fed a second 7025 that was set up as a pair of cathode followers with a lower-impedance output, as seen on the Super Twin.

Most combo bass amps generally have 100 watts or less, which is just not enough for many situations. With 200 watts of tube power into an efficient bass speaker capable of handling that power, this amp was certainly enough to keep up with all but the loudest of drummers. Add elaborate tone controls, a compact size, and a reasonable price, and you should have a hit, whether it's 1995 or 1980. Perhaps this amp will join the reissue series someday.

Attempting to "cover all basses" with one amp, Fender came up with a winning combination of power and volume as well as manageable size and high-fidelity tone. An Electro-Voice 15" coupled to a 200-watt all-tube amp put out enough deep bass for almost any playing situation.

75 1980–1982
30 1980–1981
140 1980

Prior to the 1980 release of the short-lived 75, 30, and very short-lived 140, Fender's methods of controlling distortion consisted of the Master Volume on the traditional amps and the Distortion control on the Super Twin and 300 PS behemoths. None of these amps had the high gain and infinite sustain that players had discovered by using distortion boxes or the all-tube overdrive of the increasingly popular Boogie amp and master volume Marshalls, etc. A new series of amps was designed with controlled multiple gain stages as well as a traditional Fender clean sound. A foot switch allowed the player to change from the Rhythm, or clean, channel to the Lead, or overdrive, channel without moving his hands from the guitar. These amps were Fender's first real venture into channel switching, as well as the "modern" sound and response of massive distortion.

The 75 was the most popular and long-lived ('80–'82) of the three. Silicon rectifiers, a 12AT7 *long-tailed* phase inverter, and two 6L6GCs were nothing new, although 75 watts from a pair of these tubes, as Fender claimed, certainly was. The blackface control panel with white numbers was a tip of the hat to earlier traditions, but the rest of the amp was unlike any of the preceding models, save for the reverb section (12AT7 driver, 7025 recovery). Five stages of 7025 were used for the preamps.

Front-panel controls were Bright switch, Volume (clean), Treble with pull-knob Boost, Middle with pull-knob Boost, Bass with pull-knob Boost,

A manageable size and a high-gain, channel-switching circuit offered an alternative sound to the traditional Fender line. Early models had a black grille with white plastic trim, as seen on all the new releases of the late seventies. This would change to the silver sparkle grille of the regular Fender line.

Lead Drive, Reverb, Lead Level, and Master Volume. The Standby and AC on/off switches were moved to the front panel, more convenient for the player but a potential source of noise.

An interesting feature was the High/Low Power switch, which cut the plate voltage from 500V to 250V and the fixed bias from 56V to −23V, decreasing the power to a manageable 15 watts. Speaker options included 1x15 and 1x12 combos and 4x10 and 2x12 piggybacks.

A shorter-lived version, the Fender 30, actually had two discrete channels, with the Normal channel having two inputs, Bright switch, Volume, Treble, and Bass, and the Reverb channel having two inputs, a Channel (select) switch, Preamp Gain with pull-knob Boost, Treble with pull-knob Boost, Middle with pull-knob Boost, Bass with pull-knob Boost, Reverb, and Volume. A 5U4 rectifier tube was something of a throwback for Fender designers to include at the time, but the rest of the electronics in the Normal channel were equally traditional: two stages of 7025 preamp, a 12AT7 phase inverter, and two 6L6GC power tubes. The Output Tubes Matching pot of most Fender amps of the time was replaced with the old-style bias adjust pot. Three stages of 7025 preamp were used for the Reverb channel and all its controls. It's interesting that Fender used a pair of 6L6s in both the 30 and the 75, which seem to have been created by two different design teams trying to accomplish the same thing: a high-distortion, channel-switching combo.

A third amp in this series was the 1980 piggyback 140 head and 4x12 cabinet. Costing $200 more for the head than the 75 head and $150 more than the Dual Showman head, this was the amp that replaced the 300 PS as Fender's top-of-the-line guitar amp. The 140 was not around for long, and little information appears to have been printed. It was the first of the three new amps to be discontinued, not long after its debut.

FEATURES

- New Hi/Lo Power Switch provides option of 75 watts RMS (165 watts Music Power), or ± 15 watts RMS tube sound.
- Rhythm channel for traditional Fender Sounds. Lead Channel for New Sounds.
- New Dual Pedal Switch with 20 foot cable, controls Lead/Rhythm and Reverb ON/OFF selections. Red lamp indicates lead "ON". Green lamp indicates Reverb "ON".
- New Six (6) Spring Reverb unit to enhance Reverb channel Rhythm tones.
- Master Control to simultaneously adjust Rhythm and Lead volume.
- Lead Level Control to adjust Lead Volume
- New Treble Boost Switch on Treble control.
- New Middle Boost Switch on Middle Control.
- New Bass Boost Switch on Bass Control.
- New Effects In/Out Jack.
- Bright Switch.
- Extension Speaker Jack.
- Line/Recording Jack.
- Output Tubes Matching control.
- Hum Balance Control.
- 3-position Ground Switch.
- A.C. Accessory Outlet.
- Jeweled Pilot Light.

The short-lived 30 was released around the same time as the 75 and shared the same objective of a medium-powered high-gain, channel-switching amp. It's interesting that the 30 only put out half the power from the same tubes.

All these amps were outfitted with the black grille and white trim seen on the Super Twin, 300 PS, and Studio Bass amps. By the time the '82 catalog was issued, only the 75 was left, and it had changed to the silver-grille blackface style. By the fall of '82, the "II" series amps replaced the 75 as Fender's answer to eighties guitar sound.

Little information exists on the 140—but with this schematic, you can build one for yourself!

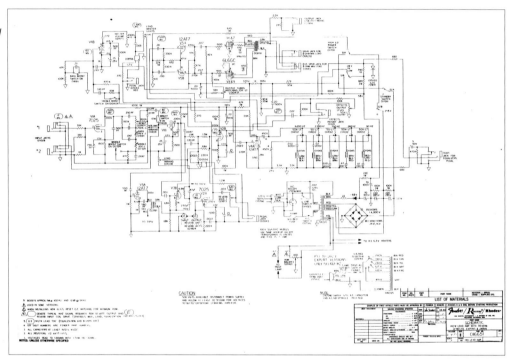

RGP-1 AND RPW-1
1982–1984

Nineteen eighty-two saw the retirement of the old-style tremolo-equipped amps and their replacement with the new "II" series. With the Twin Reverb II as Fender's only 100-watt tube amp, a modern amp was needed at the top of the line. Rack-mounted studio equipment was all the rage for guitarists, and Fender had already dabbled in this with the solid-state B-300 bass head. So the RGP-1 and RPW-1 were born.

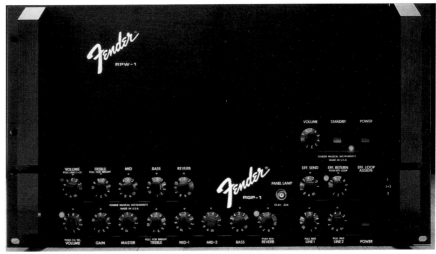

The R&D department must have sunk barrels of money and time into developing a component system to compete in the rack-mountable recording studio/coliseum rock/all-in-one-package market. At the time, this looked like the wave of the future. A dozen years later, the combo amp is more popular than ever.

An all-tube 100-watt power amp, the RPW-1 featured a balanced low-impedance XLR input (and parallel output), a pair of high-impedance inputs (why two for a mono amp?), ¼" speaker and extension speaker jacks (binding posts would have been nice), a Line/Recording Out, an Output Matching pot, and a Damping switch (Lo-Hi), which "optimizes the sound for guitar, bass, or other use." The four-6L6 amp was rated 100 watts at 1% distortion, a considerably lower specification than Fender's usual 5%. The heavy-duty case was four rack spaces tall by 8" deep.

Accompanying the power amp was the eight-tube RGP-1 preamp. This had two independent channels and a multi-function foot switch. The classic Fender channel was fitted with Volume, Treble with pull-knob Bright, Mid, Bass, and Reverb controls. A pull-knob control on the Volume pot put both channels on simultaneously. The second channel had three gain controls (Volume, Gain, Master), Treble with pull-knob Bright, Mid-1, Mid-2, Bass, and Reverb. Pull-knob *pads* (cuts) were built into the output controls of the individual channels. An assignable effects loop (1, 2, 1+2) had front-panel controls for Send and Return. While having these on the front panel was very user-friendly, having to get behind the unit to plug in your guitar was not; a jack on the front would have been far more convenient. The back panel housed the balanced XLR and unbalanced ¼" Line Outs, the effects loop jacks, and a ¼" "Speaker 8 Ω Min." jack. (Could this preamp drive a speaker?) Fender offered 1x12, 2x10, 2x12, 4x10, and 4x12 speaker enclosures to go along with the rack-mount units.

A matching bass preamp was to be ready by late '83, but the July '84 price list showed the whole line in limbo; the guitar preamp, the bass preamp, and the power amp all had "TBA" (To Be Announced) in the price column. By the end of the year they were history. These were the last tube amps from the CBS design team.

TWEED SERIES
1993–Present

Pro Junior
Blues Junior
Blues Deluxe
Blues DeVille
Bronco

This mid-priced series of amps, introduced in '93, offers the sound of an all-tube amp to those who cannot afford, or don't require, the features of Fender's top lines. With Custom Amp Shop models over $2k, and the Reissue and Pro Tube series amps all over $1k (except the $920 '65 Deluxe Reverb), there was still a huge market in need of an affordable tube amp. The Pro Junior, a great start for under $350, is actually similar to a tweed Harvard, having just a Volume and a Tone control. The controls are between two

TWEED.
The Fender amps of the 50's didn't wear tweed because it looked great. No, tweed was the choice fabric for luggage of the era. Tough enough for travel, tough enough for the road was the thought. And you'll find the Tweed Series to be tough enough, too. Classic bluesy sound and modern features will take you to new levels. And vintage styling will take you back to the good old days.

Although the solid-state Bronco does not fit in with this section, the remaining three, which were joined by the Blues Junior in '95, are all practical, low-tech tube amps with basic features and mid-line prices.

sides of a 12AX7 twin triode; a second 12AX7 is used for the phase inverter, and—new for Fender—a pair of 6BQ5s (EL84s) serve as the power tubes. A single input and the Volume and Tone controls are mounted, along with an On/Off switch and a fuse holder, to a top-mounted chrome control panel. With a *narrow-panel* tweed cabinet, pointer knobs, appropriate grille cloth, and a 10" speaker, this is a very enticing package either as an entry-level amp (in the black Tolex Princeton tradition) or as a minimal-circuitry recording amp waiting to be cranked past its 15W rating. A fancier version with a 12" speaker, Reverb, a Master Volume with foot-switchable "Fat" switch, and separate Bass, Mid, and Treble controls joined the series in '95. The Blues Junior, as it is called, offers a host of features not found on the Pro Junior at a mere $100 premium.

For an additional "buck and a half," the Blues Deluxe offers two channels (Vintage and Drive), an effects loop, and 40 watts from two 6L6s. A 12AX7 phase inverter and three stages of 12AX7 preamp (a fourth is grounded out, unused) complete the tube complement. A fourth model, the 60W Blues DeVille, is available as a 4x10 or 2x12 combo. Same tubes as the Blues Deluxe, but higher voltages.

All but the Pro Junior use solid-state devices for the reverb, effects loop, foot switch, and rectifier functions, with all-tube preamp and power amp sections (signal path). The Pro Junior, except for its solid-state rectifier and printed circuit board, could pass for a fifties circuit. The entire line became available in white Tolex with maroon grille cloth in '95, an interesting look for the fifties-style boxes.

The new Bronco, while counted as part of the Tweed series, is actually a solid-state amp (see page 154).

CUSTOM AMP SHOP
1993–Present

Vibro King
Tonemaster

Yesterday's amps today, with a few additions. The look is classic 1961 Fender; you can't fault the designers for their choice, as these have long been considered the "coolest of the cool." Point-to-point wiring, taken for granted up until 10 years ago, is today a major selling point, as is the inclusion of a tube Reverb Unit in the Vibro King. Tube tremolo, working the fixed bias of the two 6L6 power tubes (like a '63 brown Vibroverb), is another big plus. Otherwise, this is a fairly simple design (especially compared to the Concert, Super, and Twin models that constitute the Professional Tube series). Volume, Treble, Mid, and Bass are the only other controls. A foot-switchable Fat switch gives a small boost and a bit more edge to the non-Master Volume circuit, which powers three 10" Alnico speakers, *à la* late-fifties Bandmasters.

Custom Amp Shop series, '93–. *Built using the traditional hand-wired, point-to-point method, these amps seem somewhat old-fashioned compared to the goings-on inside the Twin-Amp and its predecessor, The Twin. Tone reigns over features.*

The Tonemaster head is available for those that require a high-gain amp. Channel switching between independent Vintage and Drive channels with separate tone controls, effects loops, and volumes makes these practical for live performance. The matching cabinets (two or four 12s) also suggest stage use. These two amps were the first in a series to come out of the Custom Amp Shop. Hand-wired, non-assembly-line work.

Rumble Bass

Rumble Bass, '94–. *Imposing to the eye and the ear.*

Deep bass...let's face it, many of today's popular bass amps simply don't have it—the "feel" portion of the bass-frequency package. The others may "cut" and offer one-hand transportability, but they can't stand up to a big bad tube bass amp, as witnessed by the continued success of Ampeg's SVT. From the Custom Amp Shop comes Fender's successor to the throne, following past royalty such as the Dual Showman, the 4x10 Bassman, and the late-forties Pro, all state-of-the-art for bass in their day. (The 400 PS Bass and 300 PS were outcasts of sorts in their day.) Three hundred watts RMS tube power and the GIANT transformers necessary to generate that much current at those voltages! Two channels each have Volume, Treble, Mid (with Cut switch), and Bass controls, plus there's an effects loop with a Mix control, allowing traditional series connections or paralleling the dry signal with the effected signal, as done in recording studios. An interesting feature is the capability of the two normally independent channels to operate in parallel, allowing one input to be affected by every knob on the amp.

Two separate 4x10 cabinets are offered, with one of each being the suggested team. For the bottom, a 4x10 SUB enclosure handling 600 watts, and for the top, a standard 4x10 with a bullet tweeter for all you snappers and poppers. No crackles from this 600W cabinet. Look for these on the big stages across the country. ($3,450 is a lot of $100 weekends at the VFW.)

Dual Professional

Top of the line...and at a time when Fender is offering more amps in more sizes and styles than ever before. Similar to the Vibro King, but with twice the power and two 12s. Dual selectable Volumes, each having its own Fat switch, can be activated by the foot switch, included free with each Dual Professional. Not cheap at a penny under three grand, but, like the Rumble Bass Amp, expect to see these on the big stages.

Prosonic

The latest from the Custom Amp Shop is the Prosonic—an amp that never was, covered in colors and materials that never were. In addition to the eye-catching appeal of the chicken-head knobs and lizard-skin covering (Sea Foam Green, anyone?), this amp has some sonic pleasures to offer: Two foot-switchable channels (Vintage and Drive), the latter of which offers dual cascading Gain controls for tailoring your overdrive *just so*; tube-driven reverb; and a three-way rectifier switch that includes the option of Class-A/cathode-bias operation, *à la* Vox AC30. Fun times lie ahead for this line!

Prosonic, '95–. *This amp boasts switchable Class-A/cathode-bias operation, for that "British" sound.*

NEW VINTAGE SERIES
1995–Present

"Custom" Vibrolux Reverb
"Custom" Vibrasonic
"Custom" Tweed Reverb

This could be the future for Fender, combining features and fashions from different eras to make new amps that are instantly recognizable but aren't what your first look suggests. Anything, no matter how anachronistic it may seem, is possible. The early models in this series, the "Custom" Vibrolux Reverb (in early-'63 white/wheat/white attire), the "Custom" Vibrasonic (in classic mid-sixties blackface), and the "Custom" Tweed Reverb (in, you guessed it, fifties tweed) are all looked at here in the chapters devoted to their namesakes.

SOLID-STATE AMPS

FIRST-SERIES SOLID-STATE
1966–1971

Dual Showman
Twin Reverb
Bassman
Super Reverb
Pro Reverb
Vibrolux Reverb
Deluxe Reverb

CBS Fender released its first series of solid-state amps in the summer of '66. *Fender Facts* #12 (September '66) featured transistorized versions of three favorite Fender amplifiers: the Dual Showman, the Twin Reverb, and the Bassman. Also featured were a Solid-State Reverb Unit and a Solid-State Public-Address System.

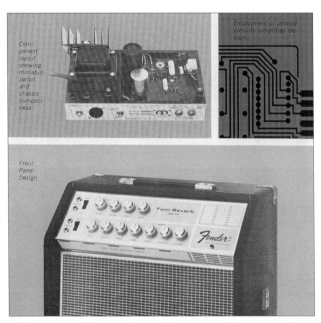

Solid-State amplifiers. *Fender in the late sixties, like everyone else at the time, thought modern technology held the answers to the world's problems.*

The three amps all were rated at 100 watts RMS, but differed in their speaker configurations and features. The Dual Showman was a piggyback style, with two 15" JBLs. The Bassman was also a piggyback, with a 3x12 cabinet. The Twin was a combo, with two 12" speakers mounted one above the other. The Showman and Twin Reverb heads were identical, offering two channels, reverb, tremolo, and a knob that matched the others but was connected to a three-position switch instead of a potentiometer. This "Style" switch controlled three different settings: "Pop," "Normal," and "CW/RR." The Bassman had a four-position Style control, with "Bass Boost 1," "Bass Boost 2," "Guitar Normal," and "Guitar Bright."

The control panels were silver, with cylindrical flat-topped knobs that were unique to this series. The speaker cabinets were trimmed in stylish aluminum. This trim would become standard on all the Fender amps. A raised nameplate, with tail, graced the early models. These were replaced with a flat piece of trim mounted horizontally across the top of the cabinets and having the words "SOLID STATE" printed boldly in red.

Fender Facts #14 (July '67) announced four new solid-state combos: the 4x10 50-watt Super Reverb, the 2x12 50-watt Pro Reverb, the 2x10 35-watt Vibrolux Reverb, and the 1x12 25-watt Deluxe Reverb. The Super Reverb and Pro Reverb featured the Style switch seen on the big amps; the Vibrolux and Deluxe were set up more like their tube counterparts.

The Dual Showman was the first solid-state amp to go, with the release of the Super Showman, not long after the release of the small combos. In 1969 all the first-series solid-state amps were retired except for the Bassman, which stayed in the line until 1971, when the whole solid-state project was called off.

The names sound familiar.... *Adding insult to injury, the solid-state amps assumed the names of the tube models. Looks like they missed the plane—and the boat.*

SUPER SHOWMAN
1969–1971

Replacing the solid-state Dual Showman in 1969 were the totally new Super Showman XFL-1000 and XFL-2000. The name Super Showman referred to the "brain" only, a three-channel preamp. The XFL-1000 was a powered speaker cabinet, holding four 12" speakers and two 70-watt power amps, putting out 140 watts total. The package as offered by Fender for $1,495 was a Super Showman head and two 4x12 cabinets, putting out 280 watts. Remember that bands were expected to fill the venue they were playing with their instrument amps; P.A. systems were for vocals. Fender's idea was that more powered cabinets could be added, at $550 apiece, until satisfactory coverage was obtained.

Super Showman XFL-1000 *(right)* **and XFL-2000.** *Designed by the legendary Seth Lover, of Gibson "P.A.F." fame, the innovative Super Showman amps proved too big and expensive to last.*

Theoretically, a power amp should be as close physically to the speakers as feasible, and the amplifier-to-speaker cable as short as possible. So powered speakers make good sense. This amp could have changed history, as powered speakers have become popular today in Hi-Fi systems.

The XFL-2000 had the same power as the 1000, but was loaded with eight 10" speakers. Two bottoms with a head cost $1695, and extra cabinets were $650 each, unless ordered with JBLs, which cost $1000 per cabinet!

The head had a built-in E tuner—a trendy item at the time—and a Master Volume that controlled the combined output of the three individual preamps, which were connected in parallel internally. The first channel had Volume, Bass, Mid, Treble, and Fuzz, a first for Fender. The second channel had Dimension IV in place of the Fuzz. This was a chorus-vibrato-underwater effect that Fender also produced as a separate effect unit (see page 170). The third channel had the traditional complement of Reverb and Tremolo.

Due to the high cost, it's doubtful that anyone but professional players used these amps, saving the everyone-plugs-into-one-amp bands that were popular at the time from fighting over channels. ("You had the Dimension IV *last* night....")

For the lone user, Fender recommended running jumpers from channel 1 to 2 and from 2 to 3, dividing the output of the instrument among the three channels. The volume control for each channel would then control the amount of signal from that channel which would be blended into the total sound of the preamp. These channels were completely independent of each other, unlike the usual series connection of plugging the output of one effect box into the input of the next. The intensities of the effects could be preset by controls on the back of the preamp marked Reverb Accent, Sound Expander Accent, and Fuzz Accent and used in conjunction with an elaborately wired foot switch.

Channel patching. *In a precursor to channel switching, short patch cords were made available so the player could access the effects of all three channels at once.*

Each powered speaker cabinet had a volume control on the back, in case you needed to have one cabinet a different volume from the other(s). Also on the back was a "Basic Sound" switch, which offered the choice of "Solid State" or "Vacuum Tube." (I wish I knew how they accomplished that!) The cabinets were connected to the brain with a special stereo cable; nothing else would work.

This fascinating amplifier was quite complex, even by today's standards, and was the result of a great deal of research and development. It's doubtful Fender had recouped its R&D costs by the time the Super Showman was phased out with the rest of the solid-state amps in 1971.

ZODIAC AMPS
1969–1971

Libra
Capricorn
Scorpio
Taurus

Fender's moon must have been in the nut house when they designed this series. With the Zodiac amps they put what they had learned from the first series of solid-state amps (They should have learned to stay away from them!) into an alligator-skinned package that offered nothing new but the name and the cosmetics. The Libra and the Capricorn came with four and three 12" JBL speakers, respectively, and were equipped with the same basic setup as the earlier solid-state amps: 105 watts, two channels, with reverb and tremolo. The Scorpio came with two 12" JBLs and a 56-watt head, and the Taurus had two 10" JBLs and 42 watts.

The Zodiacs were offered for under two years and probably were only manufactured for a very short run, as they are quite rare today. Collectible? Everything is collectible, but who cares? (If they had made one for my sign, Aquarius, I'd certainly be tempted!) It would be another 10 years before Fender would again offer a transistorized guitar amp.

B-300
1980–1982

In 1980 Fender replaced the all-tube 300 PS Bass head with the new solid-state B-300 head, weighing in at less than 30 lbs. This was the company's first transistor amp in almost 10 years, and it was *expensive*—$1660 with the Deluxe Bass Enclosure loaded with two 15" Electro-Voice speakers.

While the 300 PS Bass had circuitry designed in the early seventies, the B-300 was definitely an eighties amp. Its features—rack-mountable, built-in compressor and variable crossover, balanced line out, foot-switchable quasi-parametric EQ, effects loop with level control, and 300 watts into 4 ohms—made this the ultimate bass amp, even by today's standards.

Assuming the amp was reliable and sounded good, one must blame its short run on economics; it simply cost too much for most of the public. It also may have been too technologically advanced for the average bass player of the time, as many of its features, considered common today, were on the cutting edge at the time.

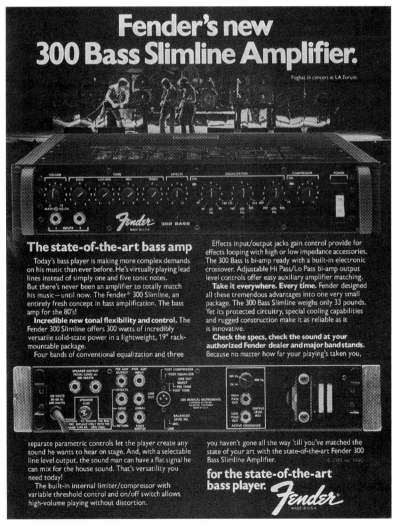

SECOND-SERIES SOLID-STATE
1981–1987

Harvard
Harvard Reverb
Bassman Compact
Sidekick Bass 30, 50
Bassman 120
Harvard Reverb II
Yale Reverb
Studio Lead
Stage Lead
Montreux
London Reverb
Showman
Sidekick 10, 20, 30

After 10 years without a solid-state amp, Fender snuck a couple of small, taller-than-wide ones in with its tube line. The $189 Harvard and the $239 Harvard Reverb were aimed at the student market. For a short time both the 6-watt 1x8 tube Champ (and Vibro Champ) and the 20-watt 1x10 solid-state Harvard (and Harvard Reverb) were available, offering the choice of a bare-bones tube amp or a feature-filled transistor amp for about the same cost. The Harvards had built-in distortion with their Master Volume controls and offered a Headphone Out and a Line Out on the rear panel. A Speaker Off switch was just one more reason to buy one of these to go with your kid's new Bullet guitar. The Champ was soon stepped up to a Super Champ (by '83), and the bottom of the line has been occupied by solid-state amps ever since.

Harvard, '81–'82. Fender avoided drawing attention to the solid-state circuitry in these amps.

A Bassman Compact was released around the same time as the Harvards, featuring a 15" speaker and 50 transistor-generated watts. A built-in compressor, Preamp Out and Power Amp In jacks, and a Master Volume were some of the newer features. The Bassman Compact was discontinued along with the rest of the bass amps in '83, leaving Fender with two new Sidekick Bass amps—the 30 and 50—and the announcement of a new solid-state Bassman 120 (1x15, 100W) for 1984. These would be the last CBS bass amps.

PART II: THE AMPS—*SECOND-SERIES SOLID-STATE*

A new line of solid-state guitar amps was released in '82, including a new Harvard Reverb II. This amp was shaped more like the retired Champs and was equipped with three tone controls and three gain stages (Volume, Gain, Master) for maximum distortion. The expanded solid-state line was mixed right in with the tube amps for the '83 amps brochure and included the new Yale Reverb (1x12, 50 W), Studio Lead (1x12, 50W), Stage Lead (2x12 or 1x12, 100W), Montreux (1x12, 100W), London Reverb (100W head, 1x12 or 2x10), and Showman (1x12, 2x12, 2x10, or 1x15, 200W). These were joined by the Sidekick 10 (1x8, 10W), 20 (1x10, 20W), and 30 (1x12, 30W). (The Sidekick 10 was designed to run on DC from a car battery or an optional battery pack.) These were sold into 1987.

Sidekick amps, '83–'94. *The Sidekicks formed a bridge from the second- to the third-series solid-state lines.*

London Reverb, '83–'86. *Amp design entered the modern era with features like channel switching and graphic EQ.*

THIRD-SERIES SOLID-STATE
1986–Present

Squier 15
Sidekick Series
15, 85, Deluxe 85
Princeton Chorus
Power Chorus, Ultra Chorus
Pro 185, Stage 185, London 185
Bassman
Keyboard 60
M-80
R.A.D., H.O.T., J.A.M.
Champion 110
Princeton 112, Deluxe 112, Stage 112SE
Performer, Champ
Bullet, Bullet Reverb
Bronco
BXR Series
KXR Series

BXR 400, '87–'94. *Solid-state design has always been kinder to bass sound than to guitar, so it made sense for FMIC to inaugurate their U.S-made solid-state line with a bass amp.*

The new company that followed CBS had no amp manufacturing facility and at first relied on old stock left over from '85 and a line of imported solid-state amps (1986) that included the Squier 15 (1x8, 15W), a similarly equipped Sidekick 15 (also available with chorus), and a number of other Sidekicks: the 25 (1x10, 25W), 35 (1x12, 35W), and 65 (1x12, 65W). Sidekick bass amps included the 35 Bass (1x12, 35W), 65 Bass (1x15, 65W), and piggyback 100 Bass (1x15, 100W). In '87 three more Sidekicks joined the line: the Sidekick Switcher (1x12, 30W), Sidekick Bass (1x10, 30W), and Sidekick Keyboard (1x10, 30W). A new line of American-made bass amps, the BXR (Bass Extended Range), was announced. The first model was the 400 top, with twin 200-watt amps, 11 bands of EQ, and a compressor, with a choice of 1x15, 4x10, or 1x18/2x10 cabinets.

Nineteen eighty-eight saw the release of a whole line of new American-made solid-state amps, a number of which were given the names of classic Fender tube amps, unused at the time. The Deluxe 85 (1x12, 85W), the Princeton Chorus (2x10, 2x25W), and the Pro 185 (2x12, 185W) had nothing in common with their namesakes except their pirated names. Others in the series were the 15 (1x8, 15W), 85 (1x12, 85W), Stage 185 (1x12, 185W), and London 185 top (185W). New bass amps included the Bassman (1x15, 60W) and the piggyback BXR 100 (1x15, 100W). The new Keyboard 60 (1x12/1x4½, 60W) offered a full-range amp, and the Sidekick Switcher, 35, 65, and 100 Bass were discontinued.

In 1989 the gray-carpet-covered M-80 (1x12, 90W) and the Power Chorus (2x12, 2x65W) made their debut. In '90 the M-80 Chorus (2x12, 2x65W) and the similarly carpeted R.A.D. (1x8, 20W), H.O.T. (1x10, 25W), and J.A.M. (1x12, 25W) combos joined the line. These small amps were aimed at the young student and featured push buttons to activate their effects. The BXR 300C, a 300-watt bass combo with a single 15" speaker, was also added at this time.

M-80 series, '89–'94. *Like the incendiary device for which they were named, the M-80 amps were designed for "explosive" sound.*

The growing solid-state line reached a peak in '91–'92 with over 25 models. These were the last years for most of the Sidekick line, but produced the student-model R.A.D. Bass (1x10, 25W) and the M-80 Bass (1x15, 160W), both covered in carpet. The Sidekick Bass 30 (1x10, 30W) and 15B (1x8, 15W) were short-lived additions. The Sidekick 100R (2x12, 100W) made its only appearance in '92, and the hybrid Champ 25SE was added that year as well.

The remainder of the Squier and Sidekick amps were given their walking papers in '93. Solid-state additions that year were the Champion 110 (1x10, 25W), Princeton 112 (1x12, 35W), Deluxe 112 (1x12, 65W), and Stage 112SE (1x12, 160W). The Power Chorus was replaced by the Ultra Chorus, and the 85, Deluxe 85, Stage 185, and London 185 were discontinued. The BXR 100 became a combo. A pair of hybrid solid-state amps with one tube, the Performer 650

FENDER AMPS: THE FIRST FIFTY YEARS

Hybrid and solid-state amps, c. '94. *Fender has expanded its offerings in this area over the years.*

(1x12, 70W) and 1000 (1x12, 100W), were new additions, as was a stripped-down Champ 25.

In '94 the Champ 25, R.A.D. Bass, and M-80 Bass bowed out, while the BXR 60 (1x12, 60W) and the Bullet (1x8, 15W, with or without reverb) made their entrance.

As Fender approaches its fiftieth birthday, its solid-state line seems to be stabilizing after almost 10 years of constant change. The hybrid Performer series features the 650 and 1000; the standard series runs from the 15-watt Bullet to the 160-watt Stage 112, and the Champion, Princeton, and Deluxe are still in the line. Princeton Chorus and Ultra Chorus amps complete the standard series, and the R.A.D., H.O.T., and J.A.M. models are popular with the youth market. (A return of the Bronco as a solid-state amp in the otherwise all-tube "Tweed Series" is somewhat confusing to anyone trying to keep track of the different models being offered.) For bass, a selection of BXR combos—the 15, 25, 60, 100, 200, and 300—offers models in all ranges, and the KXR 100 and 200 are designed for keyboards or as self-contained full-range P.A. systems.

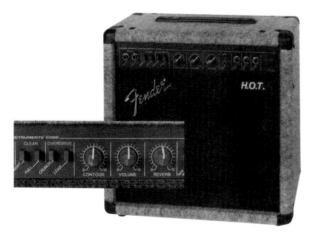

H.O.T., '90–. *The R.A.D., H.O.T., and J.A.M. models, with their push-button access to preset sounds, have gained the favor of younger players.*

EFFECTS

VOLUME PEDAL
1954–1984

Volume Pedal
Volume Tone Pedal

The Fender company has over the years offered a wide variety of outboard devices to place between the guitar and amp for manipulation and enhancement of the final sound emanating from the speaker. The first of these devices was the Volume Pedal. Gibson, Epiphone, Rickenbacker, and DeArmond had all offered a basic volume pedal before WWII, when the electric guitar was less than 10 years old. By the late forties, Fender was offering the DeArmond pedal in its catalogs and price lists. In 1954 Fender released its own version, a low-profile pedal with a metal frame and a rubber grip pad glued to the top. An input jack and an output jack were mounted on the right side as you looked down, and the throw was designed so the pedal would operate comfortably for the player in either standing or sitting position. The price was $36.50 until 1964. This pedal stayed in the Fender line until 1984 with the only change being a script logo molded into the rubber, which replaced the original block letters in the mid sixties.

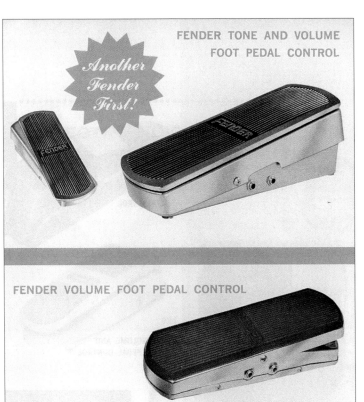

DeArmond and Bigsby had been offering a combination volume-and-tone pedal before Fender had even its Volume Pedal, and many steel players considered it a necessary part of their setup. Muted runs, ascending slides from muted to full treble (à la "Looney Toons"), imitating trains and the human voice—these were just a few of the tricks made possible with such a pedal. In 1958 Fender finally offered its version, with the standard front-to-back movement controlling the volume, while side-to-side movement turned a tone roll-off control. The rubber pad originally bore the Fender name in block letters; this was changed to a script logo around 1962. This version remained basically unchanged until it was dropped, along with the Volume Pedal, in 1984.

Although these pedals didn't generate a great deal of income for Fender, there was no real expense in keeping them in the line as a customer service; moreover, by discontinuing them, Fender left its potential customers with no choice but to buy and use a pedal with a name on it other than theirs. Oh, well....

ECCOFONIC
1958-1959

When discussing effects one must start with Les Paul and his post-war experiments using electric solidbody guitars, multitrack recording, recording at different speeds, and tape echo. The echo effect was certainly one of the most pleasing, and also the easiest for home tinkerers to copy. While Cleveland DJ the Mad Daddy was using runaway tape echo on his voice to accent words, and science fiction movies were exploiting the effect, several parties were attempting to make Mr. Polfus' studio sound, as well as the fabulous Sun Records sound, available for the everyday guitar setup. These efforts resulted in such devices as the Echosonic Amp, the Echoplex, and the EccoFonic. Fender offered the EccoFonic (which, like the DeArmond accessories, was brought in from outside the company) in '58 and '59, built into a tweed carrying case to match their amps of the time.

Here's how it worked. A loop of magnetic recording tape passes an activated record head. This puts a signal on the tape. A short time later this signal passes the playback head. (The tape speed and the physical distance between the record head and the playback head control how much time it takes for the signal that was just recorded to reach the playback head.) As the signal passes the playback head it is reproduced and electronically mixed with the original source signal.

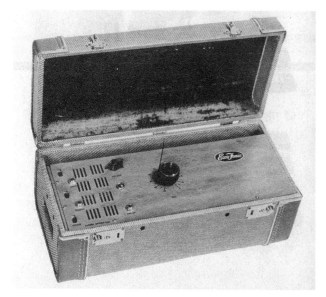

The original signal was taken from the leads to the voice coil on the speaker of the user's amplifier. An input volume control, three different delay times, and a feedback control were the variables one could experiment with.

In '59 EccoFonic began its own promotion, offering the unit *sans* tweed case. The company lasted only a few years, and it would be 1963 before another echo unit was offered by Fender.

REVERB UNIT
1961-1966
1976-1978
1994-Present

Reverb Unit
Tube Reverb
Fender Reverb
"Custom" Tweed Reverb

The most influential of Fender's effects would have to be the tube-driven spring-reverb-equipped box, referred to by the company as the Reverb Unit. According to Dick Dale, a recipient during the prototype stage, it was originally used for vocals; but after a short experiment with a guitar, it became obvious where the majority of these devices would end up. The '61 catalog showed a brown Tolex model with a leather handle and a flat logo. Instead of grille cloth, the front panel was also covered in brown Tolex. Controls for Tone, Mix, and Dwell (duration) allowed a variety of tonal moods, from brash and overwhelming to rich and supportive. Brown knobs were stock.

Prototype ('61).

The Hammond Organ Company had developed the spring reverb to enhance the sound of the electric organ. For years, this system was the only economical way to simulate reverberation, so companies like Fender who wanted to offer reverb to their customers licensed the Hammond design.

The power supply of the Reverb Unit included a solid-state *half-wave* rectifier followed by *big* filter caps and a choke to overcome the shortcomings of the center-tap-less transformer. The signal path started with a twin triode 12AT7 preamp, with the Dwell control between the two halves. Electrically this was simply an input level control for the effected signal's path, but due to the physical response of the springs, the higher the signal to the springs, the longer they would sustain a signal. A 6K6GT (similar to a 6V6) driver for the transformer and a two-spring *pan* were followed by half a 7025 as the reverb recovery tube. The output of this tube fed one leg of the mixer pot. The original input signal was split between the 12AT7 preamp (effect) and the second half of the 7025 tube (dry). This 7025 section was run as a cathode follower adding no gain (actually a small loss) and was connected to the other leg of the mixer pot. The output jack was connected directly to the wiper of this pot, sweeping from all original signal to all reverb, sounding best somewhere in between. The rubber strips on the inside of the pan were not in the circuit diagram but were an important part of the sound, damping vibrations in the pan and keeping them out of the sensitive spring circuit.

A pan lock bracket inside the box (and barely visible) shoves the springs up against foam rubber mounted to the inside of the front panel. An old trick used by dealers to beat pawn shops was to shove the unseen bracket in so that the springs couldn't vibrate. "I'll take it even though it's broken if you wanna deal on it" was usually followed by "But it was just working…how much do you want off?"

Production model ('62).

The production models had wheat grille cloth instead of Tolex for their front panels. A second color option was white Tolex with maroon grille and white knobs to match the piggybacks and the Twin. A flat logo nameplate accented the look, matching the larger cabinets of the amplifiers. The possibility exists of maroon-grilled brown Tolex units to match the last of the big combos with this combination, but they would indeed be rare birds. The possibility also exists of special-order tweed-covered units to match the majority of the amps that were already out there in '61.

As the cosmetics of the amps changed, so did the Reverb Units: white Tolex with wheat grille ('62), then black handle ('63), raised logo ('63), and finally smooth white Tolex with gold sparkle grille (late '63). The brown unit also added the black handle, the raised logo, and smooth Tolex in '63. Also in '63 was the introduction of black Tolex with silver cloth. These black units had white knobs (and black control panel with white numbers) instead of the new black-skirted numbered knobs. In the summer of '66 the new solid-state reverb replaced the tube model, although it looked nothing like the tube amps. It matched the new solid-state line, but they were all equipped with reverb; who knows what the powers-that-be were thinking.

From '76 to '78 a Tube Reverb was again available, this time with a silver panel and black knobs. It was very similar to the original, with a few differences. The power supply still relied on a half-wave rectifier, even though a new international voltage transformer was used, so they weren't just using the old circuit or using up old parts. The choke was not used, but even bigger filter caps helped smooth the pulsating DC, and a hum balance control balanced the heater voltages, reducing 60-cycle problems. Better-quality diodes were readily available by this time, and one big one replaced the original three. The dwell and tone circuits were the same, but the driver tube was changed to a more common 6V6. An extra 7025 was added to the last part of the circuit as a buffer stage, isolating the output jack from the Mixer pot. A nice addition was a wire from the output jack to the input jack that shorted the output unless something was plugged in the input. The original units made a horrendous racket if bumped while plugged into a live amp, as did the silverface models once a guitar was plugged into the input; try smacking one for a special effect!

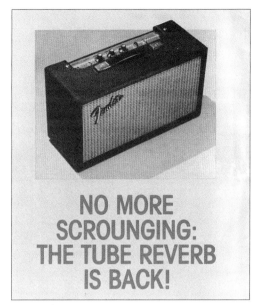

First "reissue" (c. 77).

Another reissue, more faithful to the original, was released in '94. Officially titled the Reissue '63 Fender Reverb, it is available with black, white, or brown Tolex. The signal path is virtually identical to its model of 30 years previous, although a printed-circuit board has replaced the hand-wiring of the original. A 6V6 replaces the 6K6, which is no longer in production, and the power supply has been greatly improved to a full-wave rectifier. A grounded AC plug is also new. To comply with modern electrical codes, a zener diode has been added to protect the cathode bypass cap for the 6V6 in case the tube shorts out.

Nearly 35 years after the release of its first Reverb Unit, Fender has finally offered one covered in tweed, to match the thousands of tweed amps still around from the fifties, as well as the Reissue Bassman amps. The chrome control panel and black pointer knobs separate this "Custom" Tweed Reverb from the Reissue line, as does the 25% price premium; but it sure does make a nice mate if you have a tweed amp.

"Custom" Tweed Reverb ('95).

TR 105
1961–1962

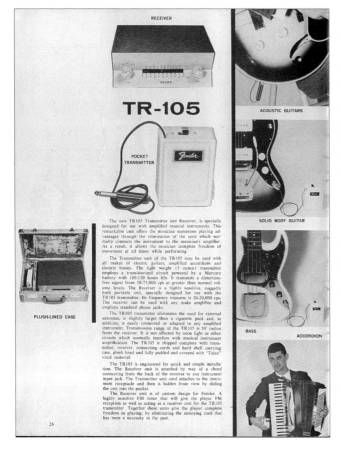

Everything old is new again! Today the wireless remote is a common part of the performing guitarist's setup. In the mid to late seventies, finicky but reasonably dependable wireless systems were beginning to be used by performers like Peter Gabriel and the members of Devo. Hard to believe Fender offered one as far back as 1961.

Had these caught on, the history of rock 'n' roll stage moves would surely read differently. On a more serious note, a number of electrocuted musicians might still be alive, and a lot of shocked hands and lips wouldn't have felt the effects of poorly grounded equipment or bad wiring. The electrical isolation a wireless gives a performer is one of the best reasons to consider one, though this probably is not what Fender had in mind when they released the TR 105. Why it lasted less than a year is a mystery; perhaps there was trouble with the FCC.

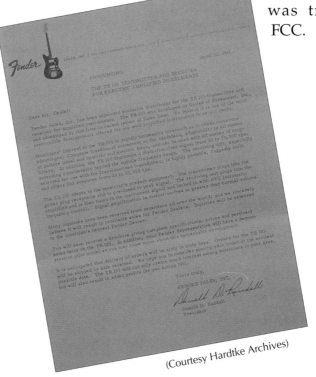

(Courtesy Hardtke Archives)

ELECTRONIC ECHO CHAMBER
1963-1968

Four years after the EccoFonic, Fender finally released its own tape echo—the Electronic Echo Chamber, as it was called. Inserted between a guitar and amplifier with a regular guitar cord, the EEC gave the musician control over Delay Time, Regeneration (number of echoes), and Intensity (echo volume). The EEC holds the honor of being Fender's first piece of transistorized processing equipment. A restyled version to match the solid-state amps was offered shortly before the unit was retired in 1968.

First version ('63).

Solid-state styling ('67).

SOLID-STATE REVERBERATION UNIT
1966–1972

Replacing the tube Reverb Unit of '61 was the 1966 solid-state Reverberation Unit. This model also employed the Hammond spring design, but used transistors instead of tubes for gain and matching. If Fender was hoping to phase out or totally restyle its tube amps, this unit would certainly support that idea. The solid-state Reverb was styled after the new solid-state amps, all of which had built-in reverb (except for the single-channel Bassman), so it obviously was not designed to be used with them. The tube amps without reverb, however, kept their blackface design for another year before changing to the silver panel with blue lettering, and would stay in the line long after the solid-state amps and Reverberation Unit were gone, so who knows? One of the suggested uses of the solid-state Reverb was with the new solid-state P.A. System, which it matched nicely; but this could not have been the primary intention for the unit. It was offered as late as 1972, bearing no resemblance to the rest of the line. Fender may have been trying to move old stock.

ECHO-REVERB
1966–1970

Echo-Reverb
Variable Echo-Reverb

Originally referred to as the Electronic Echo Chamber Disc Delay, the Echo-Reverb differed from the Electronic Echo Chamber tape delay in its method of storing sound. The standard tape loop was superseded by a rotating metal-coated disk that would be electrostatically charged (record) and spun past stationary playback heads, which would in turn send a delayed signal to be blended with the original source. The principle is the same as a tape echo except for the method of storing the signal. The Echo-Reverb was available from early '66 until late '70 (model "II" added a Bright switch somewhere during the run), but lacked some of the hands-on control associated with the popular Echoplex, which could be manipulated while playing. The three fixed delay settings (long, short, and combined) were limited when compared to the Echoplex's system of sliding heads, which could be fine-tuned to the tempo of the music or slid closer to or farther from the record head for special effects, such as pitch bending. The Echo-Reverb did feature reverb along with the echo, and also a warbling vibrato effect to fatten up one's sound, like it or not.

Echo-Reverb.

The Variable Echo-Reverb debuted in 1970 as an attempt to address the shortcomings of the original model, offering variable speed "from Very Slow to Very Rapid." Both units were offered side by side, with the Variable unit costing about 25% more than its older brother. Both were gone by year's end.

SOUNDETTE
1967–1968

A second tapeless echo device was the Soundette, which operated using a spinning magnetic drum and four fixed playback heads. The Duration control set the number of repeats, the Echo control set the blend, or mix, and the Volume control set the overall level.

Three push buttons activated three fixed delay times, but an interesting option was the ability to combine more than one delay time at once, which created a pulsating effect. Another treat of the Soundette was its tinted window, through which one could see the drum spinning (or the universe expanding, depending on one's outlook, remembering this was '67 and '68).

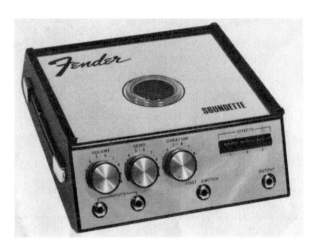

VIBRATONE
1967–1972

While not an electronic device *per se*, the Vibratone is to this day one of Fender's most useful effects. Taking the Leslie rotating speaker design (CBS owned the Leslie patents at the time) and applying it to a speaker system for guitar, Fender came up with a new item that has served guitar players noticeably better than the original.

Here are some reasons why:

First, the Vibratone was designed for use with a regular guitar amp. The Leslie usually came with a power amp, but no preamp, so the built-in amp was usually bypassed. The weight and expense of the Leslie's amp are negatives for something that won't be used.

Second, the Vibratone was built to be roadworthy—covered in Tolex, with metal corners. The Leslie was designed to be permanently installed in one's living room or church. When used in a rock 'n' roll environment, the beautiful wood cabinets quickly became chipped-up plywood, with rings in the finish from drinks set on the top.

Finally, the Vibratone used a guitar speaker, with a frequency range that guitarists were accustomed to. All but the smallest Leslies were two-way speaker systems, with a woofer, a high-frequency horn, and a crossover. This full-frequency system can sound harsh, as the highs that would normally not be reproduced come through loud and clear. Listen to a guitar plugged into a mixing console connected to full-range studio monitors and then plug it into a combo amp with a single 15" speaker. Get the picture? The two-way system also required two mics to get the full sound into a P.A. or recording console; the Vibratone needs only one. (Suggestion: try a mic on each side.)

(Courtesy Brian Fischer)

To explain how a Leslie works, the Doppler effect must be defined. This is the perceived change in pitch as a source of sound is moved closer to (raising the pitch) or farther away from (lowering the pitch) the listener. A passing train or police siren is a good example.

The rotating horn of a Leslie sends sound waves directly to a listener as it passes in front of him. As the horn goes around, the source of the sound (the opening of the horn) moves away from the listener until it has gone halfway around—at which point the sound is pointed directly away from the listener—and as it continues rotating, the sound moves toward the listener again. This causes two things to happen. First is the Doppler effect, as the repeated spinning causes the pitch to alternately fall and rise. (And as the sound is reflected off the inside of the cabinet and off of surrounding surfaces, a variety of rising and falling pitches blend together.) Second is a mild tremolo, since the direct sound is louder than the indirect sound.

Rotating baffle. *The board on the bottom is actually the back of the cabinet.* (Courtesy Brian Fischer)

Leslie cabinets generally use rotating horns, with baffles placed in front of the cone speakers, rather than rotating the speakers themselves. (A lightweight baffle is considerably easier to spin around than a 10" speaker, and creates the same effect.) On the Vibratone, the cone pointed out instead of down, and sound was expelled through ports on the sides and on top of the cabinet, slightly different from the standard Leslie, but the same general idea.

A double foot switch controlled two motor speeds and turned the Vibratone on and off. When the Vibratone was shut off, the amplifier's regular speaker was switched in; when it was turned on, a filter sent the extreme highs and lows to the amplifier's speaker, while the majority of the sound came from the Vibratone. When the amplifier's speaker was not plugged in, the Vibratone still had a full sound. The slow speed gave a rich, moody tone, while the fast gave a brighter, more cheerful sound, often associated with skating rink organs. As the baffle started speeding up or slowing down, the Doppler effect was accentuated. To hear someone well-rehearsed in using this effect is an aural delight, whether the instrument played is a guitar or a Hammond B-3.

The Vibratone was available from 1967 until 1972. Early models had a raised logo with tail; later models had the flat trim, with "Vibratone" in red.

This is a very desirable piece today because of its practicality and tone. It would be worthy of reissue.

ORCHESTRATION +
1968-1970

What was Fender thinking when they released the Orchestration + Voicing Multiplier? They were certainly hoping every trumpet and sax player in America would be enticed by the lure of effects, so that horn players could have access to all the new tones like the guitarists. Wishful thinking, because then they'd all need amps! Twenty-five years later, horns are still miked. Maybe Fender should reissue the Orchestration +.

The Orchestration + was not in any of the catalogs, but was offered in price lists from 1968 until 1970. The unit was featured in the final issue of *Fender Facts*, but two questions were left unanswered: "Where and how do I plug in my trumpet?" and "What happens if I plug a guitar into it?"

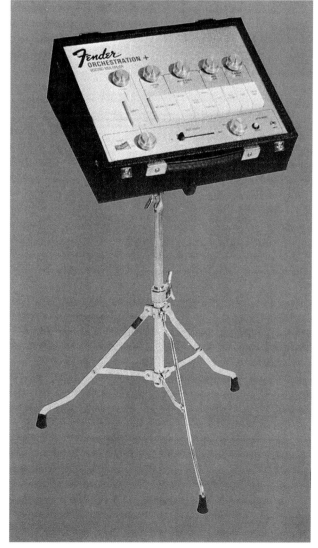

DIMENSION IV
1968-1970

Dimension IV Sound Expander
Special Effects Center

An oil-filled spinning drum provided both an underwater sound and an aura of science fiction for the appropriately named Dimension IV. These little boxes are wonderful—if they work. If they don't, leave them alone. Replacement parts are no longer available, and rumor has it that the oil can is filled with a mysterious carcinogen. If yours works, don't turn it upside down! And to keep the oil from getting so hot the effect loses intensity, don't leave the motor running.

Dimension IV. *Exterior of the Universal.*

There were two models of the Dimension IV Sound Expander: a Universal, which instruments could be plugged directly into, and for half the price, a version that connected to the Reverb In and Out jacks on the back of Fender Reverb amps. A switch on this unit gave you stock Reverb or Dimension IV and Reverb. Another turned the motor on and off, activating the Dimension IV effect. The Universal model was much larger than the other version. A plate on the back stated, "ADINEKO memory system by Tel-Ray Electronics Los Angeles Calif." Patent numbers were given—2,892,898, 3,072,543, and 3,215,911, if you're interested in learning more. Tel-Ray built a number of the sixties effects for Fender.

The Special Effects Center, introduced in 1969 and lasting into the following year, was an all-in-one unit that combined Dimension IV, fuzz, echo, and reverb. It was pricey, costing about as much as a Vibrolux Reverb amp at the time. Dimension IV was also built into one channel of the Super Showman.

The Dimension IV was available all of 1969 and part of each surrounding year. What did the Dimension IV Sound Expander do? Why, it expanded the sound, added dimension....

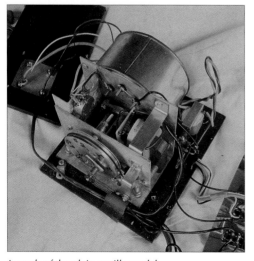

Innards of the plain vanilla model.

FUZZ-WAH
1968-1984

Unless you consider overdriven tube amps fuzz, 1968 was the year of Fender's venture into psychedelia. The 1969 catalog (assembled in '68) described the Fuzz-Wah as "far out" and "wild." This was a big change indeed.

Gibson had introduced the Maestro Fuzz in 1962, and by '68 there were any number on the market. The "Clyde McCoy and His Talking Trumpet" wah-wah effect was introduced by Vox in 1967. So Fender's was not a ground-breaking pedal. Having to operate the wah with side-to-side motion, as described in the catalog, would have seemed foreign to anyone used to the Vox. *Fender Facts* #16 (November '68) described the wah action as up-and-down, which appears to be how production models left the factory.

Around 1974 a new Fuzz-Wah Pedal was introduced; fuzz was controlled with a Fuzz Blend pot and a Fuzz Output Level pot. The wah had an output level control and a switch that turned off the wah and turned the pedal into a volume control. This version, along with all the other remaining pedals, was last seen in 1984.

First version ('68).

Second version ('76).

MULTI-ECHO
1969–1970

Like the Dimension IV, the Multi-Echo was designed to plug into the Reverb In and Out jacks of a Fender Reverb amp. This was an economy, bare-bones echo device, with connecting cables to go from the amp's reverb to the Multi-Echo, and then a four-position switch offering Off, On, Long Echo, and Short Echo. Intensity was controlled through the amp's Reverb knob. No fun to operating this. This tapeless, lifeless box lasted for the years 1969 and '70.

FENDER BLENDER
1968–1977

Joining the Fuzz-Wah in late '68 was another distortion pedal with a bit more control. An On/Off switch activated Volume, Sustain, Tone, and Blend controls, which most kids immediately turned to 10. Another switch activated a Tone Boost. This pedal ran unchanged for almost 10 years, although a later model with an octave control is rumored to have been made.

PHASER
1975–1977

The phase shifter was a seventies thing—the big Maestro, with its colorful switches; the little orange Phase 90; the Small Stone; and, for a short time, the Fender Phaser. This was a rather large pedal for only one control and an On/Off switch, but that was so you could vary the phaser speed with your foot while you played. AC-powered with a short power cord and a short life. Phased out with the Fender Blender in 1977.

MISCELLANEOUS STOMP BOXES
1986-1987

For a short time Fender was having its good name put on a series of generic imported stomp boxes that included a distortion, a flanger, a stereo chorus, a compressor, and a digital delay. They're mentioned here for posterity even though they weren't really Fender. (Then again, neither were many of the effects from the sixties.)

FENDER AMPS FAMILY PORTRAIT

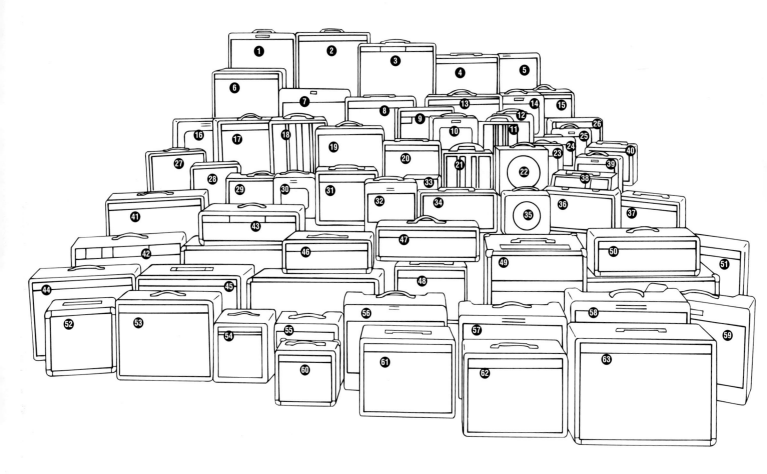

1. '59 Bassman Reissue 1995
2. Vibro King 1995
3. Vibroverb 1963
4. Stage 112SE 1994
5. Blues Deluxe.................1994
6. Concert.........................1962
7. Super 1950
8. Deluxe 1964
9. Super Champ Deluxe.....1984
10. Princeton.......................1950
11. Princeton.......................1947
12. Princeton.......................1953
13. Deluxe Reverb...............1994
14. Deluxe 1951
15. Princeton.......................1963
16. Deluxe 1948
17. Deluxe 1963
18. Professional 1946
19. Vibrolux 1958
20. Vibro Champ 1966
21. Deluxe "Model 26" 1947
22. K&F 1945
23. Champion "600" 1949
24. Champ 1953
25. Princeton.......................1948
26. Bronco 1968
27. White 1956
28. Yale Prototype? .... c. 1955
29. Reverb Unit 1963
30. Champion "800" 1948
31. Champ 1966
32. Princeton.......................1952
33. Dimension IV 1969
34. Reverb Unit 1964
35. K&F 1945
36. Reverb Unit 1965
37. Musicmaster Bass 1974
38. Echo-Reverb 1969
39. Champ 1955
40. Champ 1964
41. Quad Reverb 1976
42. Scorpio 1970
43. Showman......................1961
44. Vibroverb 1964
45. The Twin 1989
46. Bassman 1966
47. Showman......................1967
48. Bassman 20 1983
49. Vibratone 1970
50. Tonemaster 1994
51. Super Reverb................1965
52. H.O.T. 1994
53. Vibrolux 1963
54. BXR 15 1994
55. Pro Junior.....................1994
56. Dual Professional 1947
57. Pro................................1958
58. Bassman 1959
59. Pro................................1953
60. Bullet 1994
61. Princeton Reverb 1968
62. Princeton Reverb 1966
63. Twin Reverb 1966

K&F Amps, c. 1945. *Forerunners of the Fender line*

Fender Amps, c. 1946–47.
The "woodies" were made following Doc's departure

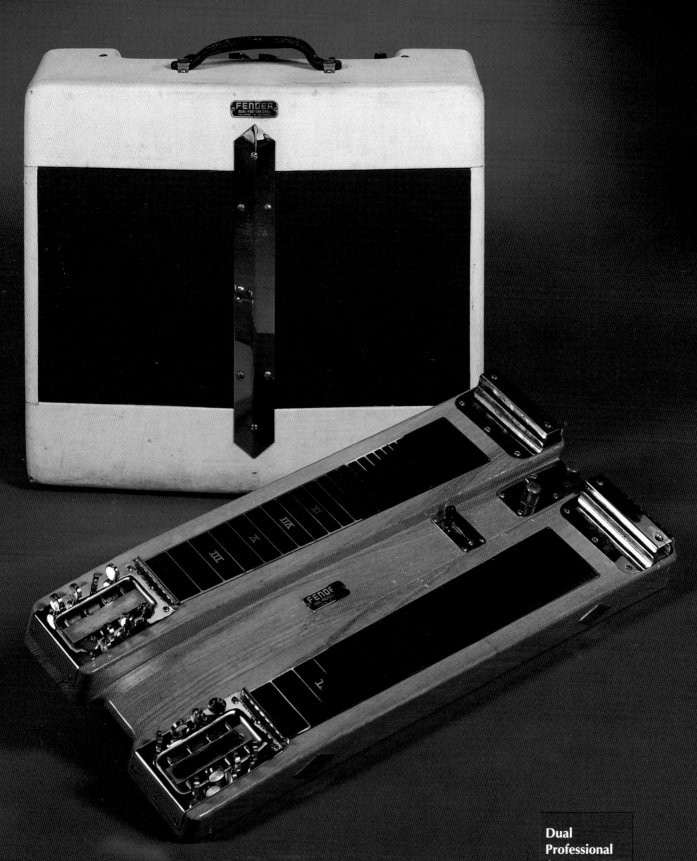

Dual Professional Amp, c. 1947. *Fender's first tweed amp*

Vertical-Tweed TV-Front Amps and Champion "800," c. 1948. *Fender's first line of tweed amps*

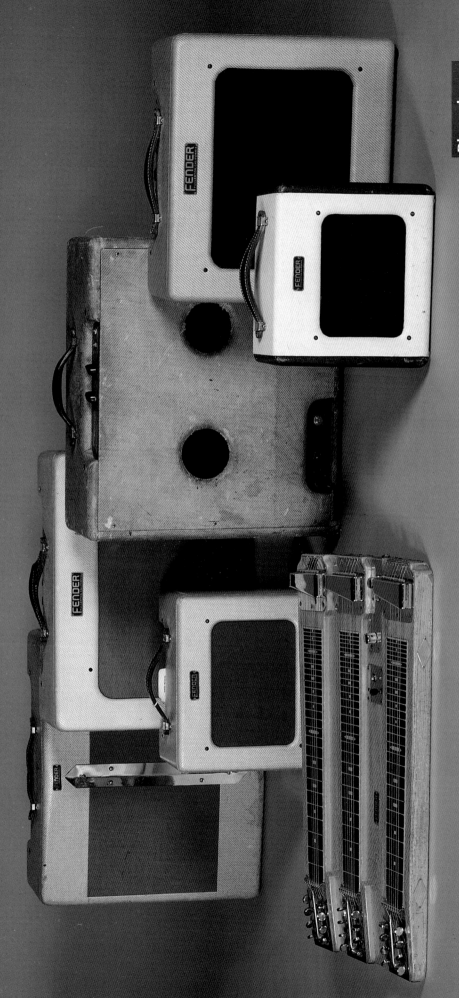

Diagonal-Tweed TV-Front Amps and Champion "600," c. 1949–52. *The diagonal tweed became a standard*

Bassman Amps, c. 1952. Early models with chassis on the bottom, in tweed and white vinyl coverings

Wide-Panel Tweed Amps, c. 1953–54. Transitional style between the TV-front and narrow-panel models

Narrow-Panel Tweed Amps, c. 1955–60.
The last of the tweeds

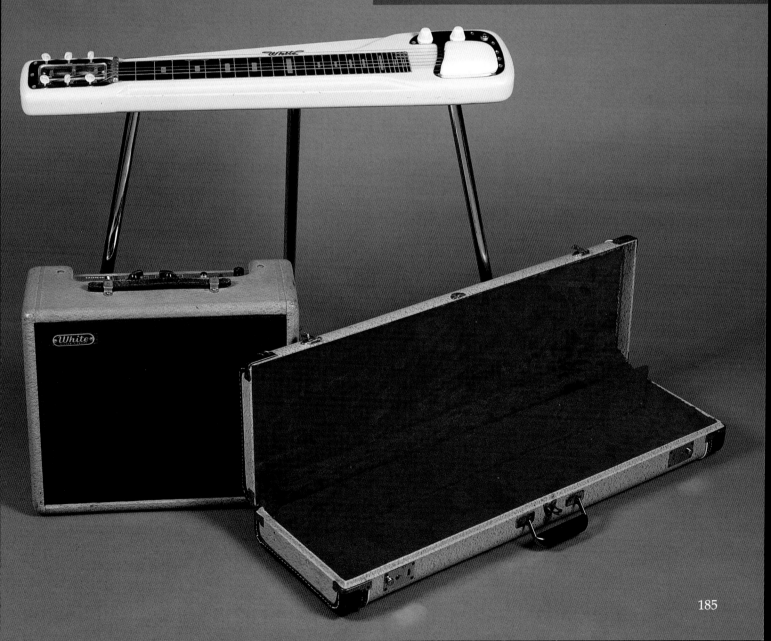

Custom-Ordered Bassman, c. 1958.
Fender's first piggyback

"White" Amp and Steel Guitar, c. 1956.
Named after Forrest White. For the student market.

Early Full-Color Flyer, c. 1958.
Featuring all 11 narrow-panel tweed amps

Cover of 1960 Catalog.
Featuring the Vibrasonic, Fender's first Tolex-covered amp

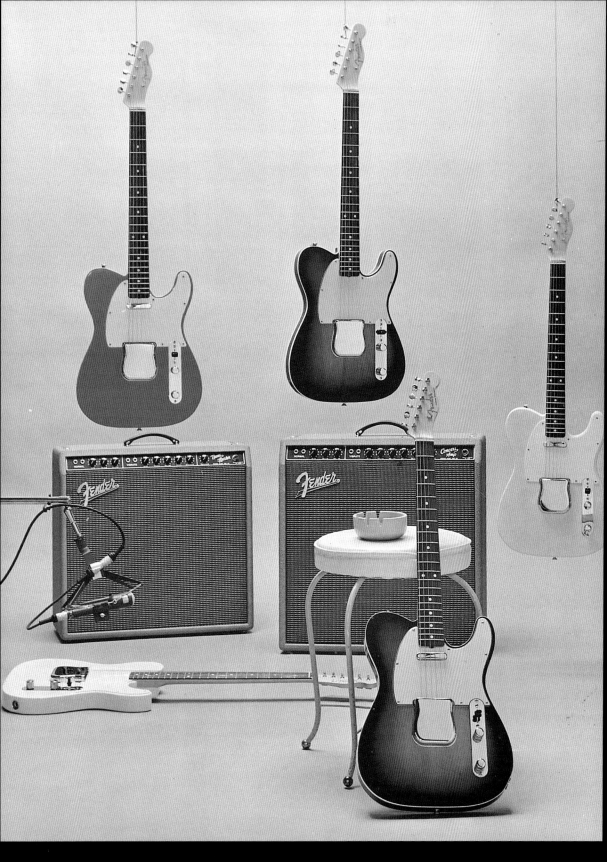

Brown Tolex Amps with Tweed-Era Grille Cloth, c. 1959–60.
Early front-mounted control panel

Rough White Tolex Amps with Maroon Grille Cloth, c. 1960–62. Early production-model piggybacks

Brown Tolex Amps with Maroon Grille Cloth, c. 1961–62

'61 Super (Missing Logo) and '60 Twin with Four 5881s. *The white knobs may not be original*

Three Styles of Brown Tolex, c. 1960–62

Brown Tolex Amps with Wheat Grille Cloth, c. 1961–63. *All the combo amps had this combination except for the Twin*

Rough White Tolex with Wheat Grille Cloth and White Knobs, c. 1962–63.
The piggyback amps also had these features

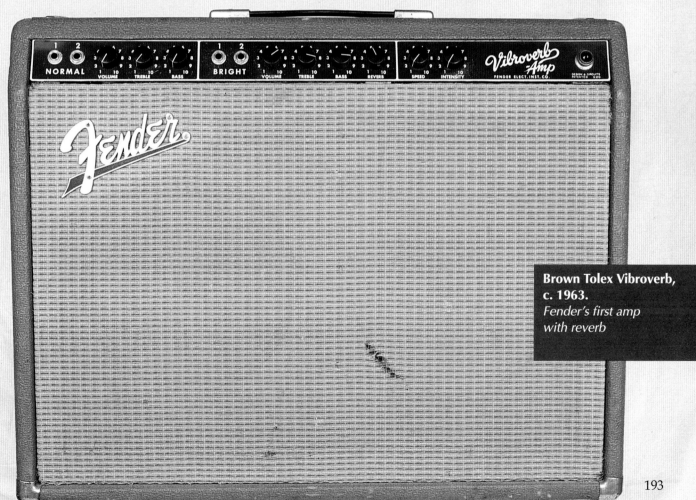

Brown Tolex Vibroverb, c. 1963.
Fender's first amp with reverb

Cover of 1963 Catalog, Featuring Black Tolex. *Notice the different grille cloth from production blackface amps*

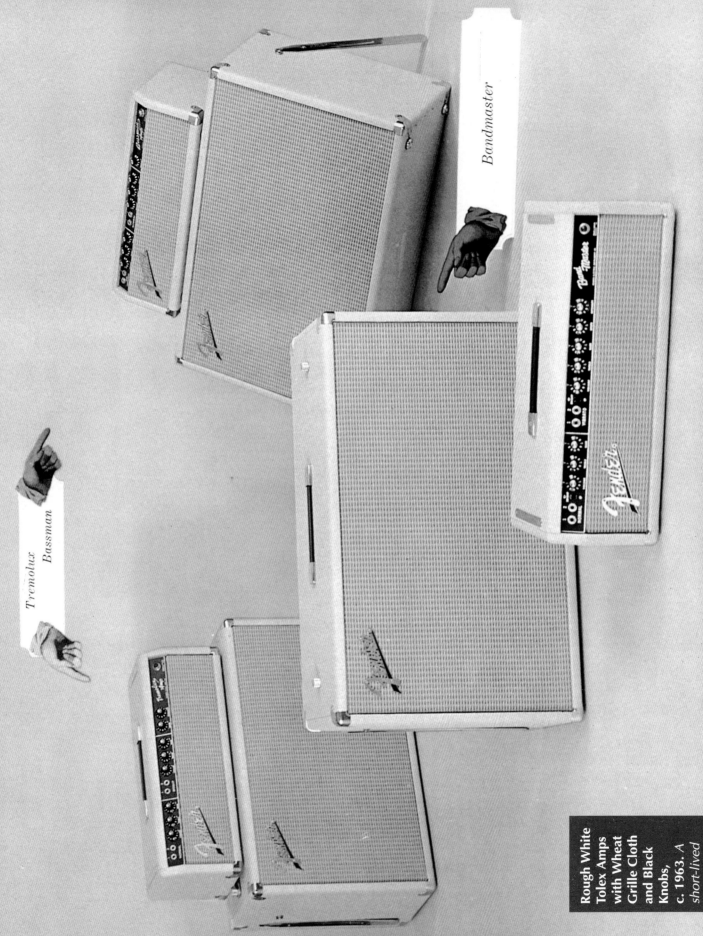

Rough White Tolex Amps with Wheat Grille Cloth and Black Knobs, c. 1963. *A short-lived variation*

Smooth White Tolex Amps with Gold Sparkle Grille Cloth and Black Knobs, c. 1964. *The last of the white Tolex amps*

Smooth White Tolex Bassman Amp with Gold Sparkle Grille Cloth and White Knobs, c. 1964. Bassmans were made with the old faceplates in '64—early ones with white Tolex, later ones with black

Blackface Combo Amps, c. 1963–67.
These amps became classics...

Blackface Piggyback Amps, c. 1963–67. ...as did these

Reverb Units, c. 1961–66. *Tube outboard reverbs to match any Tolex-covered amp*

Solid-State Amps, c. 1966–71. *Fender's early attempts to transistorize*

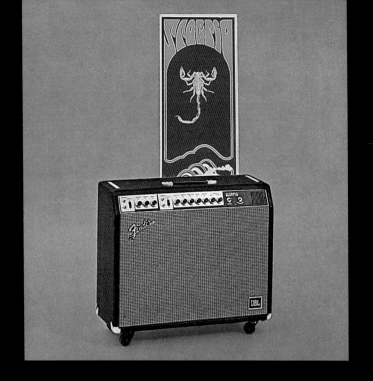

the sounds of the zodiac.

FENDER TRANSISTOR AMPLIFIERS

Zodiac Series, c. 1970–71.
The last of the solid-state amps until 1980

Silverface Amps with Aluminum Trim, c. 1967–69. *A new set of clothes for the late sixties*

Fender Amplifiers

Silverface Amps, c. 1969–81. *Fender's longest-running series*

CBS-Era Miscellany, c. 1965–85

Second-Version Blackface, c. 1982. *Short-lived cosmetics on standard amps*

"II" Series, c. 1983–85. *The last CBS-era models*

Red-Knob Series, c. 1987–92.
The first amps for the new Fender Musical Instruments

Champ 12

The Champ® 12 is truly an amplifier with something for everybody. Champ 12 features include: Two Instrument Inputs, Gain Switching, Reverb, Mid-Boost (for heavier distortion), 12" Eminence Speaker, Line Out and a Headphone Jack which automatically disconnects the speaker for private practices.

Use the footswitch to go to Overdrive mode. The tone controls shift their frequency response when switching from clean to Overdrive.

The Tape Input Jacks allow connection of a tape deck, CD player, drum machine or other audio source. The tone and distortion controls affect only the guitar sound.

Pros, students and teachers have all found something about the Champ 12 that's perfect for them.

Coverings

The Twin, Dual Showman Stack and Champ 12 are available in durable *Special* Coverings. When ordering, change the last three digits of the part number:

Color	Color Code
Black	000
Red	100
Gray	200
White	300
Snake Skin	400

Tubes

Fender engineers, world's leading manufacturers, have developed the finest low noise, high performance tubes available. Fender *Special Design Tubes* are recommended as replacements for any tube amplifier.

Fender Musical Instruments, 1130 Columbia Street, Brea, CA
© 1987 Fender Musical Instruments. Twin, Showman and Champ are registered trademarks of to change without notice.

Reissue Series, c. 1990–. *Classic sounds and looks*

Recent Expanded Line, c. 1990s. Amps for all styles and price ranges

Pro Tube and Custom Amp Shop Series, c. 1993–. *A classy combination of new and old*

Gray Crinkle, c. 1945. *K&F*

Light Wood, c. 1946–48. *Princeton, Deluxe "Model 26," Pro; available in "Gleaming Blonde Maple," as well as Oak or Ash*

Dark Wood, c. 1946–48. *Princeton, Deluxe "Model 26," Pro; "Dark Mahogany" and "Black Walnut"*

Vertical Tweed, Style 1, c. 1947. *Dual Professional; no brown line, almost white*

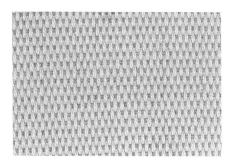
Vertical Tweed, Style 2, c. 1947–48. *Princeton, Deluxe, Pro, Dual Professional/Super; with brown line, no coating*

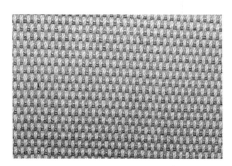
Vertical Tweed, Style 3, c. 1948. *Princeton, Deluxe, Pro, Super; like Style 2, but with yellowish coating*

Green Tweed, c. 1947–49. *Champion "800"*

Brown and White Leatherette, c. 1949–53. *Champion "600"*

Diagonal Tweed, Style 1, c. 1949–51. *Princeton, Deluxe, Pro, Super; low-contrast*

Diagonal Tweed, Style 2, c. 1952–64. *All tweed amps; high-contrast (slight variations in color and weave)*

Rough Brown Tolex, Style 1, c. 1959–61. *Professional series combo amps; note pinkish tint*

Rough White Tolex, c. 1960–63. *Piggyback amps, Twin*

Coverings

Rough Brown Tolex, Style 2, c. 1961–63. *All combo amps except Twin*

Smooth Brown Tolex, c. 1963–64. *Very last brown combos, Reverb Unit*

Smooth White Tolex, c. 1963–64. *Piggyback amps, Reverb Unit*

Black Tolex, c. 1963–85. *All amps except Zodiac series (texture varies slightly)*

Gator-Textured Vinyl, c. 1969–71. *Zodiac series*

Black Tolex, c. 1986–. *All U.S.-made amps; thick backing*

Black Tolex, c. 1986–. *Imported amps; thin backing*

White Lizard, c. 1987–90. *"Red-Knob" tube amps*

Regal Red Lizard, c. 1987–90. *"Red-Knob" tube amps*

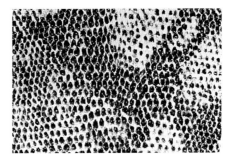
Black Python, c. 1987–90. *"Red-Knob" tube amps*

Gray Carpet, c. 1989. *M-80, RAD, HOT, JAM, HM series*

Black Carpet, c. 1993. *M-80, RAD, HOT, JAM, HM series*

Reissue Tweed, c. 1990–. *'59 Bassman, Tweed series, "Custom" Tweed Reverb*

Reissue Brown Tolex, c. 1990–. *Vibroverb, '63 Reverb Unit*

Reissue White Tolex, c. 1993–. *Custom Shop, "Vintage"-style amps*

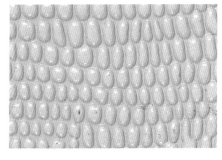
Sea Foam Green Lizard, c. 1995–. *Prosonic*

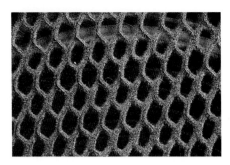
Metal Mesh with Fuzzy Coating, c. 1945. *K&F*

"Woodie"-Era Red, c. 1946–47

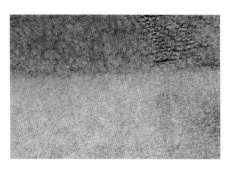
"Woodie"-Era Gold, c. 1946–47

"Woodie"-Era Blue, c. 1946–47

Brown "Mohair," c. 1947–48. *Vertical-tweed era, also early Champion "600"*

Purple, c. 1948–49. *Champion "800"*

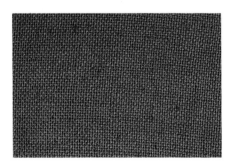
Brown Linen, c. 1949–55. *Wide-panel era (due to fading, shades vary from original dark to faded light brown)*

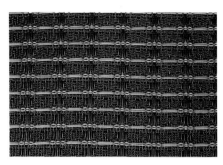
Tweed-Era Grille Cloth, c. 1954–64. *Narrow-panel tweed amps and the first brown Tolex amps; the first true grille cloth Fender used*

Maroon, c. 1960–62. *White Tolex amps and Professional series brown Tolex amps*

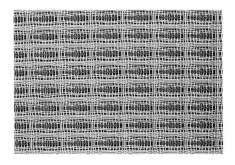
Wheat, c. 1961–63. *Small brown Tolex combos 1961–63, all amps (except Champ) c. 1962–63*

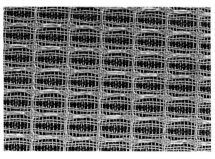
Silver Sparkle, c. 1963–67, 1981–85. *Blackface black Tolex amps (example shown is nicotine-stained; originally silver, white, and black)*

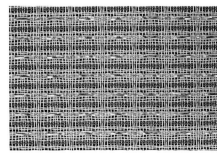
Gold Sparkle, c. 1963–64. *Smooth while Tolex amps*

Grille Cloth

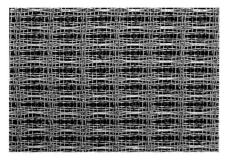

Blue Sparkle, Style 1, c. 1967–81. *Silverface amps; with blue sparkle thread (several variations exist; earliest versions trimmed in aluminum*

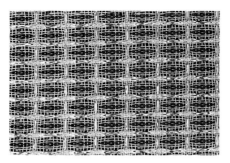

Blue Sparkle, Style 2. *Note lack of sparkle in blue thread*

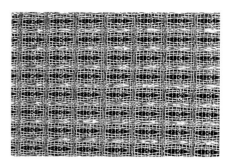

Silver and Orange, Style 1, c. 1969–71. *Zodiac series; with shiny orange thread*

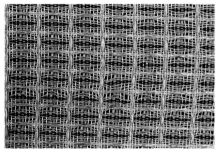

Silver and Orange, Style 2. *Many silverface amps in mid to late seventies; note lack of sparkle in orange thread*

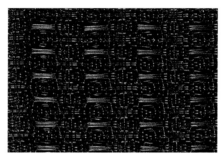

Black with White Plastic Trim, c. 1975–81. *Used on new releases during these years*

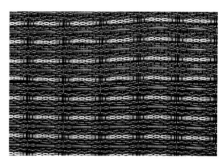

Brown, c. 1982–85. *Oak-cabinet Super Champ "Deluxe," Princeton II "Deluxe," Concert "Deluxe"*

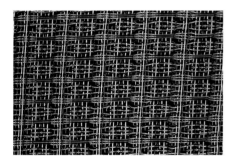

Gray, c. 1986–92. *"Red-Knob" series*

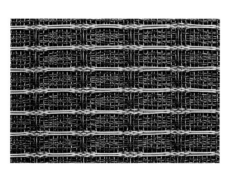

Reissue Tweed, c. 1990–. *'59 Bassman, all new tweed amps*

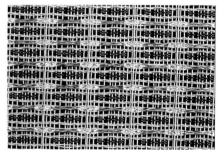

Reissue Silver Sparkle, c. 1991–. *Standard series, sixties reissues, Performer series*

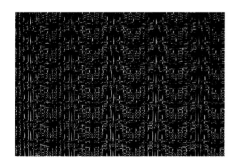

Reissue Maroon, c. 1993–. *All blonde amps except "Custom" Vibrolux Reverb and Limited Edition Deluxe Reverb*

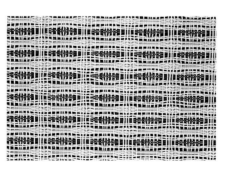

Reissue Wheat, c. 1991–. *Vibroverb, '63 Reverb Unit, "Custom" Vibrolux Reverb*

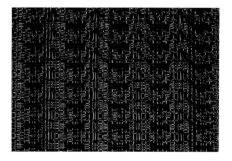

Black, c. 1989. *M-80, Bass amps, KXR Keyboard amps*

Pictures from Fender *Electric Guitar Course*, Books One and Two, c. 1966

PART IV: AMPOLOGY

AMP BASICS

The following discussion makes no attempt to explain how amplifiers work; there simply isn't the space to do that here. This section is intended merely to summarize the changes in electronics that Fender amps have undergone through the years. For complete explanations of the terms used, consult the books on electronics listed in the Bibliography.

Tubes

Fender used basically three types of tubes in its amp designs, number one being the *full-wave rectifier* type. This tube is part of the power supply, converting high-voltage AC from the power transformer to pulsating DC, which is then smoothed by the filter caps and choke. Prior to 1928, high-voltage DC had to be delivered from large storage batteries, which were labeled A, B, and C, according to their positions in the standard amplifier circuitry. The high DC voltage applied to the plates of the amplifying tubes (see below) is still referred to as the "B+" voltage. This voltage is essential for plate current to flow—the basis of operation for the power and preamp tubes.

Fender used 5U4, 5Y3, and GZ34 rectifier tubes, and on the Twin and Bassman amps of the fifties used two wired in parallel. A brief experiment was made using the 83 mercury-vapor tube c. 1957. As solid-state devices became available, the rectifier tube was often replaced with a diode bridge rectifier, made with four or more silicon diodes.

The second type of tube used is called the *power* tube. In most cases Fender used *beam* power tubes, a style of pentode (a tube containing five main

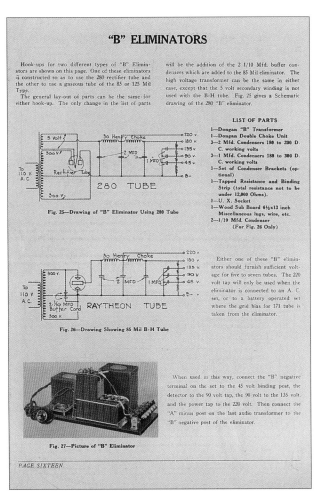

"B" battery eliminators, 1928. *The use of rectifier tubes to provide high-voltage DC was a new idea at this time.*

components). Power tubes have low amplification factors, generally under 10, but are capable of passing high currents, and therefore providing high power. Originally, Fender preferred metal-cased tubes, such as the 6V6 and 6L6. These would be replaced with the glass-bottle 6V6GT and 6L6G in the early fifties. The 5881, a high-quality 6L6G, was stock in a number of the higher-power amps by the late fifties. The 6L6GB and 6L6GC, also commonly used higher-powered versions, began to be used in the mid fifties and early sixties, respectively. In '61, the 6BQ5/EL84 was used for a very short time on the Tremolux. The 6550 was used for the 300 and 400 PS. On the 400 PS, six of these powerful tubes required a 6L6GC for a driver tube preceding the phase inverter. On the 300 PS, equipped with four 6550s, a 6V6 was used as a driver in a similar fashion. The Reverb Unit used a 6K6 driver for the reverb pan.

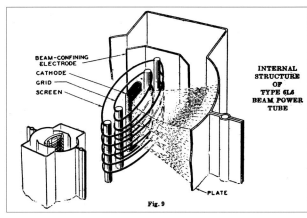

Inside a power tube. *From the* RCA Receiving Tube Manual.

The third tube type is usually referred to as a *preamp* tube, although they can be used in a number of other roles, such as phase inverters, oscillators, cathode followers, and reverb drivers. The original choice were military issue, metal-cased "medium-mu" or "hi-mu" twin triode tubes. Twin triode means that two triode circuits (cathode-grid-plate) are contained in one case. These tubes give a high amplification factor of between 35 for the 6N7 and 70 for the 6SC7. Glass-bottled 6SL7 and 6SN7 tubes were also used. All but these last two were built with a single cathode, requiring both halves to be biased the same. In the mid fifties two glass tubes, the 12AX7 high-mu twin triode (amplification factor 100) and 12AY7 medium-mu twin triode (amplification factor 44) became the standard small-signal tubes. The 12AY7 was used until 1960, when the 7025 became first choice for Fender designers. This was a high-quality 12AX7, first used in the brown Tolex amps. (Nowadays 12AX7s and 7025s are the same thing.) The 12AT7 hi-mu twin triode (amplification factor 60) began to be used in the blackface amps for phase inverters and reverb return functions. The 12AX7 and 12AT7 became the standard for Fender. In the seventies Fender joined Ampeg in the "let's use esoteric hi-fi tubes" game. The Super Twin Reverb used a 6C10 triple triode and a 6CX8 medium-mu triode/sharp-cutoff pentode. The sharp cutoff feature was built into the tube by the way it was wound and is actually a compressor, giving less amplification factor at higher input voltages. This feature was built into the first preamp tubes Fender used for their Professional and Dual Professional amps: the 6SJ7 sharp-cutoff pentode. This tube was also used in Champs until c. 1953. A 6AV6 high-mu triode (amplification factor 100) was used in the single-channel Harvard. The lower-gain 6AT6 (amplification factor 70) replaced it by '57. This was Fender's only use of a single-triode tube, although the tubes actually contained an additional circuit, a dual diode, that was not hooked up (pins 5 and 6).

Most of the tubes can still be found, if only as new old stock (NOS). In the old days, if a designer needed a special tube, chances were someone would make them. Nowadays, even with the firm plant that tubes have in the amplifier market, no real R&D for new tubes is conducted. In designing an amp, it is simply a practical design policy to use only the readily available tubes, in particular the 12AX7s, 6L6GCs, and 6V6GTs.

Bias

Fender used three basic methods of biasing tubes. The preamp tubes were usually *cathode biased*. A resistor (cathode resistor) between the cathode and ground causes a positive voltage between the cathode and ground when plate current flows. Sounds logical, right? The cathode, being positive, makes the grid negative in comparison with it. This method is also referred to as *self bias*, because it does not require an external voltage source; it creates its own when plate current flows.

For a short time Fender used the *grid resistor* method of biasing preamp tubes. (The first Fender amps used cathode-biased tubes in the preamp section, as did all the amps after '53.) The amps using the grid resistor bias method have their inputs connected directly to a capacitor and a grid resistor with a very large value (to limit grid current). Cathodes are connected directly to ground. Hum is more of a problem with grid resistor bias, a logical explanation why its use was discontinued.

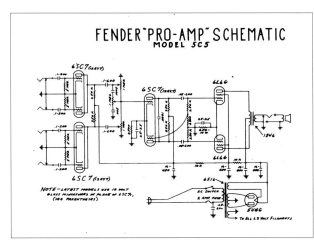

Bias. *The two 6SC7 preamp tubes at left are grid resistor biased. The two 6L6G power tubes at right are cathode biased.*

Power tubes were biased in two different manners, which give noticeably different responses when music flows through the tube. The first method is the cathode bias, as used in the preamp section. The second is known as *fixed bias*, and requires the cathode to be connected to ground and the grids to a negative voltage supply. The latter circuit was first used by Fender in the mid fifties, on all the amps but the Champ, Princeton, and Deluxe (one possible reason for the popularity of the small amps in recording studios). A separate winding on the power transformer is required for fixed bias. The alternating current from the transformer is converted to DC for the negative bias voltage by the use of, dare I say it, a solid-state device—a diode. Fixed-bias amps tend to be more powerful and have a more dynamic response than cathode-biased power amp sections, which tend to compress a bit (self leveling) and don't produce as high an output.

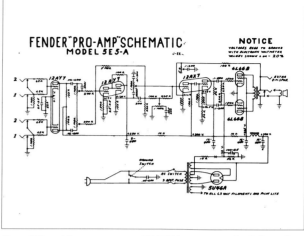

Son of Bias. *All preamp stages (12AY7s) are now cathode biased. The power tubes now have fixed bias.*

Phase Inverters

When Fender first made amps, the least-powerful amp used a single power tube to amplify the entire signal. Both the positive and the negative components of the AC were handled, as with preamp tubes. This is called *single-ended*. The larger amps were equipped with *phase inverter* circuits, so that one power tube would reproduce the positive half of the signal and another would handle the negative. This *push-pull* arrangement is much more efficient than a single-ended design, being capable of producing many times the output. Feeding the two power tubes a pair of identical but opposite signals is the job of the phase inverter. (For those who care about technological hair-splitting, the word "phase" is misused here, as it often is. What's actually involved is an inversion of *polarity*—positive becomes negative and vice versa.) One of the earliest methods required a center-tapped transformer—a relatively expensive item. The only Fenders to use this method were the 400 PS Bass, 300 PS, and Musicmaster Bass amps of the seventies. All the other phase inverters use a twin triode tube and some resistors.

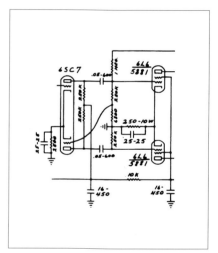

Paraphase inverter. *From Twin model 5C8. The 6SC7 is the phase inverter tube; the 6L6/5881s are the power tubes.*

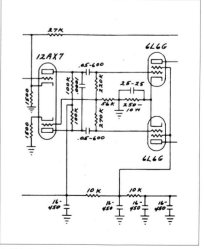

Self-balancing paraphase inverter. *From Twin model 5D8.*

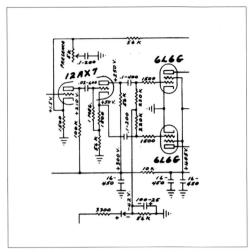

Split-load inverter. *From Twin model 5E8-A.*

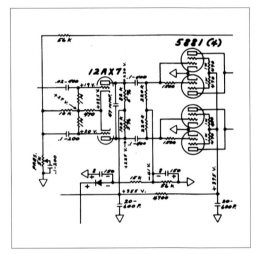

Long-tailed inverter. *From Twin model 5F8-A.*

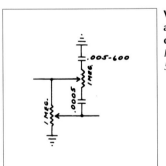

Volume *(left)* **and Tone** *(right)* **controls.** *From Princeton model 5C2.*

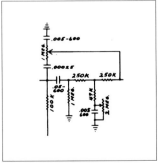

Treble *(upper left)* **and Bass** *(lower right)* **controls.** *From Twin model 5C8.*

The first circuit Fender used is referred to as a *conventional*, or *paraphase*, inverter and was used on all the push-pull amps through '53. Most of the amps c. '54 used an improved circuit, the *self-balancing paraphase* inverter. Neither of these was near ideal, and in '55 a third style was installed, requiring only half a twin triode. This *cathodyne*, or *split-load*, inverter (a.k.a. *phase splitter*) was a better design, but used a cathode follower, and therefore incurred a slight loss in level, as opposed to the extra gain provided by the first two circuits Fender used. The second half of the twin triode that was freed up (by only using one side for the inverter) added the second preamp stage necessary to push the power tubes. This design was satisfactory and continued to be used until 1960 on all the push-pull tweeds except for the last Bassmans and Twins.

A new design, the *cathode-coupled*, or *long-tailed*, inverter, was first seen (on a Fender, that is) on the big boys of the late fifties—the Twin and the Bassman—as well as the last Tremoluxes. This design was integrated at the time the Twin went to four power tubes and both amps changed to the 83 rectifier tube. It became the standard for just about all push-pull circuits, including all Fenders since 1960 (except those with transformer inverters mentioned previously), including the solid-state amps. It is still used today.

Tone Circuits

Varying the frequency response, or "tone," of an amplifier is a big part of getting a good sound. In the early days this entailed the simple roll-off of treble frequencies. The Twin introduced separate Bass and Treble controls c. '52, and the tweaking began. While these two operated on simple RC (resistor-capacitor)-to-ground networks, the new Presence control of '54 used a completely different approach. To understand this mysterious effect, we must first look at *negative feedback*, which was designed to reduce distortion and stabilize frequency response. This hi-fi trick takes a small portion of the amplified signal going to the speaker and sends it back into the amp before the power tube. As the amp gets louder, this negative feedback will tend to cancel out any distortion, unevenness in frequency response (peaks), or noise created by the amplifier.

Fender's "Presence" control acts on the negative-feedback loop by passing progressively more of the high-frequency portion to ground, leaving lows and mids to control how the feedback loop affects the rest of the circuit. As the amp is pushed, the highs start to break up; normally, the feedback loop would act to control this, but instead, it's as if there is no feedback on the highs. The intensity of the effect changes with volume. The name "Presence" was also used on the Super Twin, but this was actually an active tone boost at 3.9 kHz.

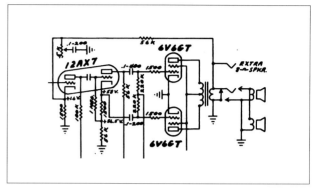

Presence control *(upper left). From Super model 5E4-A. This control operates on the negative-feedback loop.*

A fourth control was added first to the Bassman and shortly after that to the Twin c. '57. The Middle control was short-lived and was not included in the brown or white Tolex amps. It returned in '63 on the blackface Twin and Showman amps and was added to the Super Reverb in the same year. Lots of other amps have had the Middle control since, especially the progressive amps starting with the 400 PS through the Super Twin, and the 30, 75, and 140. It is still used on the "Twin-Amp" and other large amps.

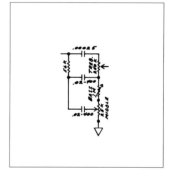

Middle control. *From Bassman model 5F6-A.*

A Bright switch was added in '63, eventually making it all the way to the Champ. Usually a simple 120pF capacitor was used, to make other makers' guitars sound like Fenders. This was replaced with a pull-knob switch on the Volume or Treble control starting around '82. (Other functions, like Cut, Boost, and Notch, have been added to other pull-knob switches hooked up to controls.)

The Super Twin, 300 PS, and Studio Bass amps of the mid seventies all used an elaborate active tone circuit that could be foot-switched in and out (see Super Twin).

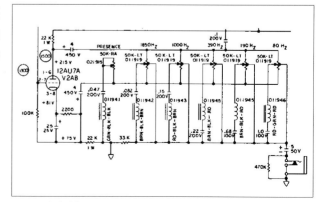

Foot-switchable active EQ. *From Studio Bass amp.*

Tremolo

Three different methods of *tremolo* (amplitude modulation; regular fluctuation in volume) have been used by Fender, and all use a tube oscillator. The Tremolux of '55 had the pulsating signal, low in both level and frequency, connect to the cathode of the phase inverter tube. This pair of triodes was cathode-biased as one, and as the signal from the oscillator "collided" with the bias voltage, the tube would be shut down and turned back on in sync with the oscillator. The speed of the oscillator and the depth, or intensity, of the pulses could be controlled with a pair of potentiometers.

A variation on this approach appeared within a few years, when the Tremolux changed to power tubes with fixed bias and used the oscillator pulses to fight against the bias supply. This is the style that was used for the brown Deluxe and Vibrolux, as well as the brown Vibroverb of '63, which inspired Fender's revival of tremolo in 1990. Working directly on the power tubes, it can be very intense.

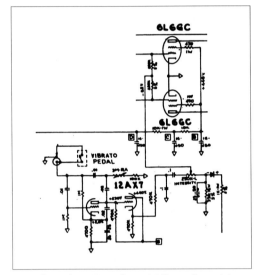

Tremolo, style 1. From (brown) Vibroverb model 6G16. The oscillator (12AX7) modulates the fixed bias of the power tubes.

A second method was designed for the '59 Vibrasonic, which was to be Fender's state-of-the-art amplifier. The earlier tremolo had a tendency to tick when there was no signal present, and it seems logical that the following design was engineered to combat this annoyance. The new design required two twin triode tubes, as opposed to only one for the Tremolux and only one half for the Vibrolux. Whereas the early method worked on the power section, the Vibrasonic's worked only on the Vibrato channel's preamp signal, post-volume and post-tone. This signal was sent to a passive crossover network that split the highs to half a 7025 twin triode and the lows to the other half. Half of the second 7025 was used as the oscillator, and the output fed the "highs" tube. It also fed the second half of its own tube, which inverted the pulsating signal. The output of this tube fed the "lows" tube. The net effect was that the tremolo made the highs pulsate against the lows—a very full and complex sound. The pulsations and inverted pulsations from the oscillator canceled at the point where the

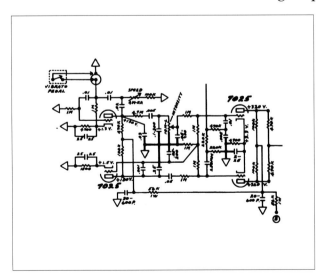

Tremolo, style 2. From Vibrasonic model 5G13. The 7025 at left is the oscillator and inverter. At right is the network that separates highs from lows, each of which is modulated separately before being summed back together.

highs and lows were summed (humbucking). Contrary to Fender's claims at the time, this is *not* vibrato, but a complex form of tremolo.

This arrangement was upgraded c. '61 with the addition of an extra 12AX7 (six small tubes instead of five). Only half of this tube was used, and it functions as a split-load inverter. The two sides of the twin triode previously used for the oscillator and inverter were now a direct-coupled cathode follower. This circuit was used on all the "Professional" amps, including the white Tolex Showman and Bandmaster piggybacks.

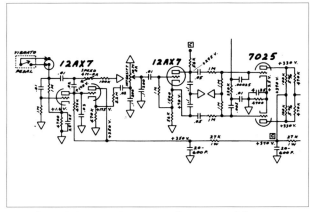

Tremolo, variation on style 2. *From Vibrasonic model 6G13-A. Fender called this "Harmonic Vibrato."*

A third major style was issued on all the blackface amps, using a completely different approach. The preamp signal was used, as with the complex circuit, but instead of the tube oscillator controlling the bias of a tube, the pulsating signal now operated a photoresistor circuit. This circuit consisted of two main parts: a small neon lamp that would vary its brightness in sync with the tube oscillator, and a light-dependent resistor placed close to the bulb; both pieces were sealed together in heat-shrink tubing. Instead of changing the gain of a tube, this new method allowed for the preamp signal to be grounded out. This method became the standard for Fender until the effect was retired, c. '82.

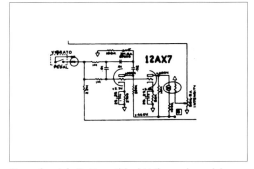

Tremolo, style 3. *From (black) Vibroverb model AA763. The egg-shaped object to the right is the photoresistor.*

Some of the new amps again include tremolo, with the reissues bringing back the first and third of the three main styles. It will be interesting to see if the complex circuit ever gets brought back. The new "Custom" Vibrasonic has tremolo operating on both channels. This would normally imply power tube bias, but the designers use the photoresistor method, applied after the two channels have been summed. Two of the three Custom Amp Shop guitar amps have tremolo, a sign that Fender Musical Instruments Corp. doesn't treat this beautiful effect as a novelty.

CBS Additions

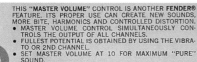

STAND BACK! NEW SWITCH BOOSTS SOUND ON AMPS

Ear-filling power for the heaviest metal sound is now available with the Fender Princeton Reverb and Deluxe Reverb Amps. The dramatic increase in sound, a thunderous 9 decibels, is available with either unit.

Designed for musicians demanding maximum power from smaller amps, the effect has been obtained by the addition of an easily operated Push/pull Power Boost switch on vibrato channel volume controls. An added plus for musicians, this feature gives a similar increase in sustain and distortion, qualities also in demand from artists requiring heavy sound.

By pulling out the volume control knob, further advantages are immediately apparent. It is now possible, with the Fender Power Boost, to obtain both a far wider range of power and have more volume available at lower settings. All fender Princeton and Deluxe Reverb amplifiers now in production will carry this feature.

One of the first external changes to the tube amps after CBS purchased Fender was the addition of the Hum Balance control, which used a potentiometer with its wiper connected to ground, and the two legs connected between the feeds for the heaters of the tubes. Originally Fender had used a center-tapped transformer section, but the center tap was removed, leading to this circuit. The Hum Balance control was a good idea, perhaps even better than the original.

The Master Volume circuit was another good idea—one that needed some time to mature. There is a *bit* of difference between "maximum distortion" and having the Master on 10. But not much, due to the circuit it controlled being "low" on high gain.

The Boost pull-knob offered welcome flexibility, but since you had to reach back and grab the Volume control, why not just turn it up? Control over the level and a foot switch would have been more practical.

MAINTENANCE AND MODS

A whole book could be written on this subject alone. Here are just a few tips:

First Steps

When you get a new amp or a "new" old amp, it's always good to do a once-over, including a tightening down of the nuts and bolts, especially those holding the speakers and transformers in place. Also, a thorough cleaning of all the pots (see later in this chapter), a new fuse (Why not? They do eventually go, and could be as old as the amp.), and a new set of your favorite tubes (A whole book could be written on this subject as well.). Have the bias adjusted by your repairman; if you don't have one, find one. Ask around; some music stores have a repair department or can recommend someone they trust. Look in the yellow pages and on bulletin boards, in the classifieds and on the Internet for established individuals and call them up, sound them out. You'll pay more for above-the-board work, but, as with many things in life, you often get what you pay for. This isn't to say that the guy who works out of his basement isn't as good; many of the best tweakers don't do it for a living and will work until things are perfect. But don't wait until your amp's broken to look for one.

The general consensus is to leave things as stock as possible, assuming you like your amp. You may have to take an old amp into the shop more often at first (don't get upset over this), but it's the "weakest link in the chain" theory: As one section is brought up to spec, another is then worked harder. Eventually the amp will stabilize. If you like the amp's sound, this is the way to go. Another approach is to replace all the caps, resistors, and tube sockets right off the bat, so that the amp is roadworthy. Even if you do this, burn it in for a while before you trust it; new parts that are prone to fail will do so in a fairly short period of time—particularly tubes; don't throw in a new set and leave for remote parts of the world. Also, don't be upset with your repairman if a tube fails soon after you get your amp home—unless you are willing to pay him to test it for a few hours at real-life volumes. Most techs won't charge you if you come *right* back. (Some people are going to hate me for that comment!)

If your amp does not have a working three-prong plug, as is the case with all amps made before the seventies, have one installed. You can save all your old gray two-prongers for collector's value, but if you're going to *use the amp*, get a grounded plug installed.

Maintenance

Once your amp is working fine, how do you keep it that way? Use it regularly, for one (seriously), which can affect capacitor life, response, and moisture content. Oxidation on connectors and tube sockets can put a spell on your amp, so reach back there and wiggle your tubes in their sockets; pull them almost out and push them back in. Unplug (and plug back in) your speaker connector a few times; this helps the contact points stay clean, assisting high-end response. The worst offenders in this category are the old RCA connectors for the reverb pan and foot switches. The metal on both the plugs and the jacks has a tendency to get a chalky look, which is oxidation, which can add noise and kill treble. Go to a hi-fi store and get a two-step contact cleaner/preserver. Once the connectors are good, twist and turn them regularly—it only takes a couple of seconds.

Pots are another point of oxidation problems. To reach these, you'll have to get inside the amp. Learn how to discharge the capacitors before doing this; there's enough electricity stored up to "kill you 'til you're dead" if you touch the wrong part accidentally. After the capacitors have been completely drained, remove the chassis, so that you can get at the back of the control panel. Turn one of the knobs all the way to one side, insert the little red tube on your can of cleaner/lubricant, and spray for a second. Turn the knob to the other side and do it again. Then work the pot back and forth a number of times. Do this for all pots, even the ones that don't get used or have their settings changed. This should be thought of as a maintenance routine, even though on some pots it may qualify as a repair. Always use a cleaner/lubricant (tuner cleaner) on potentiometers, or they will eventually freeze up. Contact cleaner is just what it says, and leaves no residue. Make sure that both are marked "safe for use with plastics."

While you have the chassis out, you can look at the caps for telltale signs—in particular, bubbles on the ends ("pregnant" caps). Often they'll still work, but it could be time to have them replaced. To look at the filter caps on most amps, you'll have to remove the housing on the bottom of the chassis, near the power transformer. Be careful!

Maintenance is an ongoing program that should be performed regularly. Don't wait until you're on stage or in the studio (where you'll really notice) to discover your amp is noisy.

Simple Mods

Once your amp is functioning cleanly and in a stable manner, you might consider a few changes. Talk to your tech! If you want it brighter or less bright, this can be accomplished by changing the values of certain components without altering the essence of the amp. Remember that many of the parts had tolerances of ±20%, so you can experiment a bit. (Not you; your repairman!) How quickly your amp breaks up and similar characteristics can be altered somewhat to suit your needs at the volumes you desire.

Don't add master volumes to amps that didn't come with them. It's 1995—do I really need to say this? If you don't like the gain characteristics after someone competent has worked with you, *get a different amp; you don't like this one!*

Tremolo speed and depth can be varied, as can the tone of your reverberant signal. Much of the bass is rolled off in this section to avoid feedback at high volumes. Depending on your other amp settings, you may be able to add more bottom confidently. (Again, not you; your repairman.) In the other direction, you can also make the treble less bright.

Speakers can make a big difference. Plug your head into other speaker setups for a test run. It's OK to change speakers in an amp you're going to use; just keep the old ones.

Replacing 6V6s with 6L6s (and a bias adjust) can make Princetons and Deluxes a bit cleaner at higher volumes. Don't expect miracles, but there's no harm in trying. *Never put a 6V6 into a socket that wants to see a 6L6.*

A lot of talk these days concerns no-master-volume silverface amps and how you can easily make them blackface in tone. Certain parts can be deleted, and bias sections can be reworked; but some of these changes were made in the first place to compensate for shortcomings brought about in other departments. Talk to your tech.

The Quest for Tone

The versatility of a Fender amp should allow you to cover a variety of styles, tones, and textures, without being pigeonholed into "the sound of the amp, like it or not." But since no one amp or tone is perfect in all situations, you need to find an amp that works for your situation and keep it in tip-top shape.

Try a number of different amps, not just the most popular models. Use them with your band, not just by themselves. (If you're not in a band, start one—even if you think you're not good enough—or if you think you're too good!) A good sound all alone often gets lost when you add drums and vocals; your sound needs to stand up in a group. This is one reason so many pros use Fenders. In a recording studio, you may find that the best sound in the final mix seems a bit clean on the individual track; and on the other hand, a track that sounds big and fat by itself may end up lost when everything else is added. This logic also applies to the stage.

If you like high-gain distortion, find a newer amp with multiple stages instead of relying on a stomp box. Just make sure your amp has a presentable clean sound. Nothing sounds worse than a delicate musical passage played through a distorted preamp section. ("Welcome to purgatory, Mr. Smith; here's your amp. Oh, yes; the gain knobs are stuck on 10!")

Learn the strengths and shortcomings of your amp. Learn where to place it and how it responds differently up on a chair as opposed to on the floor, or with its back near a wall, projecting out, as opposed to in front of the drummer, pushing sound out both the front *and* the back.

A lot of how an amp sounds is dependent on how you play. Get to know your amp and work with it—e.g., adjust the tone controls as you turn it up; adjust your guitar volume instead of always having it on 10; use less reverb when strumming chords (or more reverb when playing single notes); add just a touch of tremolo at times. Some of the best "mods" you can make to your tone don't require getting inside the amp.

PART V: APPENDICES

PARTS

Nameplates

K&F

Wooden Professional

Dual Professional

TV-Front Era

Narrow-Panel Era

Music Store Tag

Brown and White Tolex

Blackface and Early Silverface

Silverface 1

Silverface 2

Silverface 3

"Deluxe" Edition

PART V: APPENDICES—*PARTS*

Control Panels

Woodie

Vertical Tweed

Two-Tone Champion

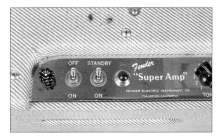
Diagonal Tweed

Brown- and Blackface

Silverface

Zodiac

Second-Version Blackface

233

Knobs

K&F

Woodie

Woodie

Tweed

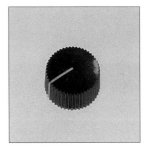
Brown Tolex

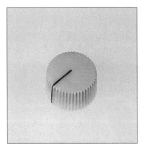
White (Occasionally Black) Tolex

Blackface

First-Series Solid-State

Zodiac

Silverface

Super Twin EQ

Late CBS

Black Pre- and Post-"Red-Knob"

"Red-Knob"

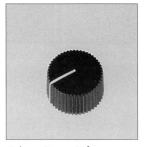
Reissue Brown Tolex

Reissue Blackface

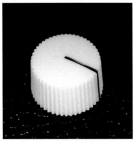
Reissue White Tolex

Reissue Pointer Knob

PART V: APPENDICES—*PARTS*

Handles

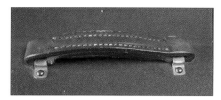
K&F

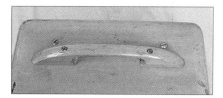
Woodie

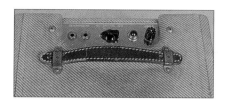
Small Tweed

Large Tweed and Small Brown Tolex

Brown and White Tolex

Blackface and Silverface

"Deluxe" Edition

Contemporary

235

FENDER AMPS: THE FIRST FIFTY YEARS

Glides

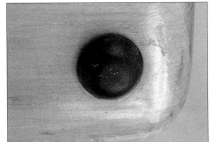
Woodie

Vertical Tweed

Diagonal Tweed

Brown Tolex

Black Tolex

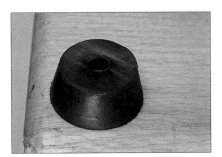
"Deluxe" Edition

New

PART V: APPENDICES—PARTS

Tube Charts

Dual Professional. *First use of a tube chart.*

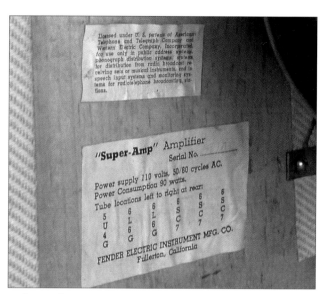

Early Tweed, c. '49. *No model number (added c. '51) or date stamp (added early '53).*

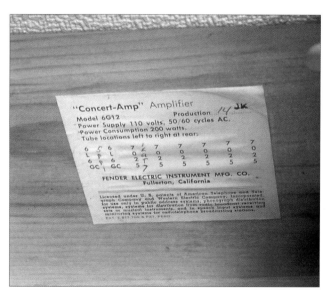

Tweed, Brown and White Tolex. *Note "JK" date stamp in upper right corner (1960, November).*

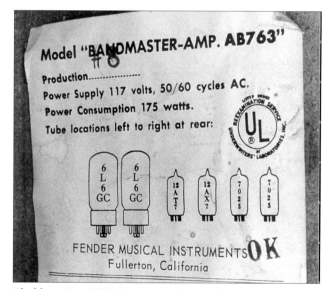

Blackface. *Note "OK" date stamp in lower right corner (1965, November). Silverface tube charts similar, but lacking date code. New-style model number (AB763) refers to schematic and circuit date (July 1963) and revision (original is AA, revised models are AB and AC). Used from introduction of blackface amps in mid '63 until early seventies.*

AVAILABILITY CHARTS, WITH PRICES

Tube Amps: Fender Electric Instrument Co.

• = amp available, but price not known
* = model change

Year	Professional	Deluxe	Deluxe Reverb	Princeton	Princeton Reverb	Super	Super Reverb	Champion "800," "600," Champ	Vibro Champ	Bassman	Twin, Twin Reverb	Bandmaster	Tremolux	Harvard	Vibrolux, Vibrolux Reverb	Vibrasonic	Concert	Showman 12	Showman 15	Double Showman	Vibroverb
1946	•	•		•																	
1947	159	77		54		159															
1948	192	84		59		159		49													
1949	192	84		59		164		*49													
1950	199	99		64		169		49													
1951	199	99		64		169		49													
1952	199	99		64		169		49		203											
1953	199	99		64		169		49		203		229									
1954	199	99		64		169		49		203	249	239									
1955	264	129		79		224		*59		339	249	289	199								
1956	264	129		79		224		59		339	339	289	199	99							
1957	264	129		79		224		59		339	339	289	199	99	129						
1958	264	129		79		224		59		339	399	289	199	99	139						
1959	264	129		79		224		59		339	399	289	199	99	139	479					
1960	284	136		82		244		62		339	399	309	209	104	149	479	359				
1961	284	139		89		244		62		349	429	329	249	104	179	479	359	550	600		
1962	289	149		99		279		66		379	429	369	279		199	479	359	550	600	800	
1963	299	169	229	109		279	389	62		389	*469	389	329		229	479	369	625	660	915	329
1964	299	169	229	109	169	279	399	69	79	399	469	389	329		*299	479	369	625	660	915	339

PART V: APPENDICES—AVAILABILITY CHARTS

Tube Amps: Fender Musical Instruments (CBS)

TBA = price To Be Announced
* = model change

The amp factory was closed with the sale of the company from CBS in early 1985. The new Fender Musical Instruments Corp. sold off old stock (or had outside subcontractors assemble amps from remaining parts inventory) into 1987.

Year	Pro Reverb	Deluxe	Deluxe Reverb, Deluxe Reverb II	Princeton	Princeton Reverb, Princeton Reverb II	Super Reverb	Champ	Vibro Champ	Champ II	Super Champ	Bassman, Bassman 50, Bassman 70	Super Bassman, Bassman 100, Bassman 135	Bassman 10	Twin Reverb, Twin Reverb II	Bandmaster	Bandmaster Reverb	Tremolux	Vibrolux Reverb	Vibrosonic Reverb	Concert	Showman 12	Showman 15	Dual Showman	Dual Showman Reverb	Bronco	Bantam Bass	Musicmaster Bass	400 PS Bass	Super Six Reverb	Quad Reverb	Super Twin, Super Twin Reverb	300 PS	Studio Bass	75	30	140	Bassman 20	RGP-1	RPW-1
1965	379	169	239	109	169	399	69	79			410			479	399		329	299		369	625	660	915																
1966	399	169	249	114	174	420	69	79			430			499	420		329	320			625	690	950																
1967	399		249	114	174	420	69	79			430			499	430			320				690	950																
1968	399		259	119	189	429	69	89			489			499	439	489		329				690	950	1000	TBA														
1969	439		289	159	219	479	74	94			539	950		549	439	539		359						1000	89														
1970	550		390	215	295	595	99	129			675	990		650	550	675		450	645					1255	94	350													
1971	460		305	167	229	500	78	99			565	835		575	460	565		379						1050	129	435	139	1250											
1972	460		305	167	229	500	78	99			*499	*599	399	575	460	565		379	645					955	99	367	139	1250											
1973	460		305	167	229	500	89	99			499	599	399	575	460	565		379						955	99		139	1250	650	675									
1974	490		340	190	260	530	100	110			550	660	425	610	500	600		410	680					1030	110		160	1250											
1975	510		350	195	270	550	105	115			575	690	440	635		625		425	680					1000			170		650	675	745								
1976	510		350	195	270	550	105	115			575	690	440	635		625		425	680					1000			170		720	750	745	1175							
1977	535		370	215	290	585	125	135			605	720	465	670		660		445	715					1000			187		720	750	745	1240	760						
1978	575		400	235	310	630	135	145			*655	*780	510	725		715		480	780					1160			210		785	790	*785	1385	820						
1979	620		435	255	335	680	145	155			705	845	550	785		775		520	845					1255			225		825	860	850	1455	885						
1980	685		495		379	755	159	169			815	905	610	875		855		575	935					1395			249		920		920		985	825	675				
1981	710		510		389	785	165	175	219	299	845	940	620	910				605	935	599							249				995			860	675	1685			
1982	710		*519		*399	785	165	175	249	349	845	940	620	910				605		649							249							860			249	599	399
1983			549		449					399	845	940		*899						759																	299	599	399
1984			599		499					399				989						759																			
1985			599		499					399				989						759																			
1986			599		499									989						759																			
1987																																							

239

FENDER AMPS: THE FIRST FIFTY YEARS

Tube Amps: Fender Musical Instruments Corp.

TBA = price To Be Announced
* = model change

The amp factory was closed with the purchase of the company from CBS in early 1985. The new Fender Musical Instruments Corp. didn't begin production of new models until 1987.

	Tube Amps							Hybrid Amps				Reissue Series					Tweed Series				New Vintage Series			Custom Amp Shop			
	The Twin, Twin-Amp	Champ 12	Dual Showman, Dual Showman Reverb	Super 60, Super	Super 112	Super 210	Concert	Performer 650	Performer 1000	Champ 25	Champ 25SE	'59 Bassman	'63 Vibroverb	'65 Twin Reverb	'65 Deluxe Reverb	'63 Reverb Unit	Bronco	Pro Junior	Blues Deluxe	Blues DeVille	Blues Junior	"Custom" Vibrolux Reverb	"Custom" Vibrasonic	"Custom" Tweed Reverb	Vibro King	Tonemaster	Rumble Bass	Dual Professional	Prosonic
1987	999	299	1249																										
1988	1199	329	1399	649																									
1989	1199	329	1399	649																									
1990	1249	339	*1749	669	729																								
1991	1299	349	1649	699	749	779						949	899	1199															
1992	1299	349	1649	699	749	779						1049	999	1199															
1993	1359		1649	*TBA			TBA	TBA	TBA	TBA	479	1049	999	1259															
1994	1399			1229			1019	489	589	399	499	1099	1059	1299	919	459	209	329	579	889					2499	1999			
1995	*1499			1239			1029	499	599	409	509	1129	1079	1299	929	469	209	339	609	899	TBA	1199	1499	599	2549	2029			
												1149	1099	1299											2599	2049	1999	2999	TBA

240

Solid-State Amps: Fender Musical Instruments (CBS)

TBA = price To Be Announced
* = model change

The amp factory was closed with the sale of the company from CBS in early 1985. The new Fender Musical Instruments Corp. sold off old stock (or had outside subcontractors assemble amps from remaining parts inventory) into 1987.

Year	Dual Showman	Twin Reverb	Bassman	Pro Reverb	Super Reverb	Vibrolux Reverb	Deluxe Reverb	Super Showman XFL-1000	Super Showman XFL-2000	Libra	Capricorn	Scorpio	Taurus	B-300	Bassman Compact	Harvard	Harvard Reverb, Harvard Reverb II	Yale Reverb	Studio Lead	Stage Lead, Stage Lead II	Montreux	London Reverb	Showman	Sidekick 10	Sidekick 20	Sidekick 30	Bassman 120	Sidekick 30 Bass	Sidekick 50 Bass
1966	1015	579	595																										
1967	1015	579	595	449																									
1968		579	595	449	499	429	359																						
1969			595					1495	1695																				
1970			745					1495	1695	899	799	650	550																
1971			625					1885	2125	1124	999	812	687																
1972								1569	1780	945	840	682	577																
1973																													
1974																													
1975																													
1976																													
1977																													
1978																													
1979																													
1980														1660															
1981														1690	395	189	239												
1982														1690	395	189	239												
1983															395		*249	349	449	499	599	699	799						
1984																	289	399	499	549	649	749	849	129	199	249	TBA	245	335
1985																	289	399	499	549	649	749	849	129	199	249	549	245	335
1986																				*439	649	749	849						
1987																				439									

FENDER AMPS: THE FIRST FIFTY YEARS

Solid-State Amps: Fender Musical Instruments Corp.

Year	Squier 15	Sidekick 15, Sidekick 15 Reverb	Sidekick Chorus	Sidekick 25 Reverb	Sidekick 35 Reverb	Sidekick 65 Reverb	Sidekick Switcher	SK Chorus 20	SKX 15R	SKX 25R	SKX 35R	SKX 65R	SKX 100R	15, X-15	85	Deluxe 85	Stage 185	London 185	Pro 185	Princeton Stereo Chorus	Power Chorus, Ultra Chorus	M-80	M-80 Chorus	R.A.D.	H.O.T.	J.A.M.	Champion 110	Deluxe 112, 112 Plus	Stage 112SE	Princeton 112, 112 Plus	Bullet	Bullet Reverb
1986	99	119	189	209	239																											
1987	99	119	189	229	239	349																										
1988	99	*139		229	239	349	269																									
1989		149		239			299													439												
1990		159		249				249	159	249	299			119	349	399	499	449	549	479	699	399	649	179	249	289						
1991	99			249				249	159	249	299	349	499	119	389	439	599	549	649	499	729	399	699	179	259	299						
1992	99													124	399	449	619	579	669	499	749	449	699	179	259	299	239	399	499			
1993														129	399	469	619	579	669	499	749	449	699	189	269	319	259	429	529	319	159	189
1994														*139	399	469	619		679	519	*649	459	719	199	279	329	259	*449	549	339	159	189
1995																			689	539	679	469		199	279	329	259			*349		

*= model change

Bass and Keyboard Amps: Fender Musical Instruments Corp.

Year	Sidekick 35 Bass	Sidekick 65 Bass	Sidekick 100 Bass (Fender 100 Bass)	Sidekick Bass (SK Bass 30)	Sidekick Bassman (Fender Bassman [100])	SK 15B	BXR Dual Bass 400 Head	BXR 300R Head	BXR 100	BXR 60	BXR 25	BXR 15	BXR 200	M-80 Bass	R.A.D. Bass	Sidekick Keyboard (SK Keyboard 30)	Sidekick (Fender) Keyboard 60	KXR 100	KXR 200
1986	239	339	499																
1987	239	329	499	199			699									199			
1988			529	219	339		699									219			
1989			529	229	349		799									229			
1990			559	239	349		829	529								239			
1991			589	249	369	149	899	599						549	249	249			
1992			589	249	369	149	899	599						599	249	249	329		
1993							899	629	469					599	259		339		
1994							899	639	499	399	289	199		619	279		349		799
1995								759	509	409			659				369	509	

PART V: APPENDICES—AVAILABILITY CHARTS

Effects

• = available, but price not known

Year	Volume Pedal	Volume Tone Pedal	Eccofonic	Reverb Unit (Tube)	TR 105	Electronic Echo Chamber	Solid-State Reverberation Unit	Echo-Reverb	Variable Echo-Reverb	Soundette	Vibratone	Orchestration +	Dimension IV	Dimension IV Universal	Special Effects Center	Fuzz-Wah	Multi-Echo	Fender Blender	Phaser	Distortion Pedal	Flanger Pedal	Stereo Chorus Pedal	Compressor Pedal	Digital Delay Pedal
1954	36																							
1955	36																							
1956	36																							
1957	36																							
1958	36	49	•																					
1959	36	49	•																					
1960	36	49			269																			
1961	36	49		129	269																			
1962	36	49		129		229																		
1963	36	49		129		229																		
1964	36	55		147		229																		
1965	44	55		149		229																		
1966	44	55		149		229	159	249																
1967	44	55				229	159	234		239	229													
1968	44	55				219	169	199		199	249	299	99	195		89		59						
1969	49	59					189	229			269	299	99	199	349	99	99	59						
1970	67	80					260	240	399		365	405	135	275	475	135	135	80						
1971	52	62					199				285					105		62						
1972	52	62					199				285					105		62						
1973	52	62														105		62						
1974	55	70														110		75						
1975	55	70														110		75	130					
1976	55	70		200												110		75	130					
1977	60	75		210												115		75	130					
1978	65	80		210												125								
1979	75	90														140								
1980	85	100														150								
1981	85	100														150								
1982	90	105														158								
1983	90	105														158								
1984	90	105														158								
1985																								
1986																				75	95	95	75	299
1987																								299

243

DATING YOUR FENDER AMP

The most important reason for dating an amp is to place it within a group of similar amps. For instance, a '64 Vibroverb would be a blackface 1x15 like all the others of that year. A '63 Vibroverb could be a blackface 1x15 or a brown 2x10—two completely different animals. A '55 Bandmaster is either a 1x15 or a 3x10—again, completely different animals. A 3x10 from '55 has less in common with a 1x15 made only a few months earlier than it does with a 3x10 made in '59. So it's not how old something is, it's the model that counts. For that reason, model numbers were included in the text (e.g., Deluxe model 5C3).

Don't be surprised if two "matching" amps sound completely different. The first wide-panel amps are closer in circuit design to TV-front models, and the later ones are closer to narrow-panel models, so there can be a big difference between what seems to be two of the same amp, even though production of wide-panels lasted only two years. And remember, older is not necessarily better.

Let's start in the middle of the Fender chronology, because that's where most of the "hip" vintage amps are from, and that's where the accuracy is best. All amps from the late forties on came with a tube chart glued to the inside of the cabinet—a simple piece of paper with a list of the tubes used and, by c. 1951, the model number. By May of '53 a two-letter code was hand-stamped in ink somewhere on the chart. The first letter designates the year, starting with C, for '53, through Q, for '67. The second letter designates the month, starting with A, for January, through L, for December. The size of the letters and their placement vary, as do their clarity. When they are blurred, logic can aid in deciphering certain cases. (For example, a white-knobbed Bandmaster would have to be '61, '62, or '63, so the first letter would have to be K, L, or M.) The author has seen these date stamps as late as QL (December '67) on blackface amps. No silverface amps could be found with the hand-stamped code.

A number of methods help date the pre-1953 and post-1967 models. A code (exposed by Hans Moust, of the Netherlands) is stamped into the top or side of all the pots. The same code is painted onto the frame of speakers. These codes usually have six digits. The first three designate the manufacturer (e.g., 137xxx is CTS, 220xxx is Jensen, 465xxx is Oxford, 328xxx is Utah, etc.). The number following the manufacturer represents the year. Initially this was a single digit—the last digit of the year (e.g., 2205xx could be from '55 or '65). In the sixties an extra digit was added to clarify the decade (e.g., 13776xx is from '76). The last two digits are the week. Thus 220209 is a Jensen speaker from the ninth week of 1952 or 1962. Unfortunately, parts can be changed, so *always* doubt the pot codes. Another reason for skepticism is that no one knows how long the parts sat in warehouses and parts bins before being used.

The silverface amps can be split into four groups:

1. Silverface with aluminum trim. Call it '68 if there are no codes to dispel that (e.g., a speaker dated 1969).

2. Silverface with no master volume. Call it early seventies—that's close enough.

3. Silverface with master volume, regular power. Call it mid seventies.

4. Silverface with master volume, high power (70 and 135W versions). Call it late seventies. Again, look for date codes on parts.

The second-version blackface amps were only around for '81 and '82, and the "II" series amps from mid '82 till '85, though some were assembled after that. (There may be a way to date these from serial numbers—does anybody know?)

Amps by the new company (funny how it's still called "the new company" 10 years after the fact) have serial numbers starting with "LO" (Lake Oswego) that seem to run in sequence; no information was available from Fender. The individual chapters in Part II of this book should help you at least get an idea of the era in which your amp was made, and for most folks, that should be enough. If you were born after '67 and you want an amp from the year you were born,

I'm afraid I can't help you. Someday a serial number list from the CBS years may turn up, FMIC may release theirs, or someone will compile one. Until then, enjoy your amp of unknown year.

For amps made before May of '53 there's hope from serial numbers and the tube charts. The "model number," which is a combination of a chronological code and the actual model number, seems to have begun appearing on tube charts sometime in '51 with the code 5Ax, the last digit being the actual model number (e.g., Deluxe 5A3). Model numbers went from smallest to largest in the early amp line: 1—Champ; 2—Princeton; 3—Deluxe; 4—Super; 5—Pro. Thereafter, new numbers were added as they were needed: 6—Bassman; 7—Bandmaster; 8—Twin; 9—Tremolux; 10—Harvard; 11—Vibrolux; 12—Concert; 13—Vibrasonic; 14—Showman; 15—Reverb Unit; 16—Vibroverb. The first two characters in the "model number" apparently changed to 5B for '52, 5C for '53, 5D for '54, and 5E for '55. After '55 they no longer changed yearly. They apparently related to circuit designs and schematic numbers. (No schematics can be found with model numbers earlier than 5Cx, but the actual amps definitely had 5Ax and 5Bx labels.) Only use these as approximate, because they may represent the year of the new circuit (rather than the release of the new model), which in the early fifties apparently changed just about every year. An early wide-panel Deluxe with model number 5B3 and a speaker with a date code of the fifty-second week of '52 supports this theory, as do numerous other examples.

For the years 1946 through 1951 there's *some* hope. A wooden-cabinet Princeton, Deluxe, or Professional would have to be '46, '47, or '48, so call yours a '47 and you won't be off by more than a year. Not good enough? The Princeton apparently did not have either a serial number or a suggested retail price (only wholesale prices were shown in the price list; the Deluxe and Professional showed retail and wholesale—40% off retail). It certainly did not have a control panel. The Deluxe "Model 26" used a control panel with the serial numbers stamped in. These appear to run in consecutive order starting at least by #219, the earliest found and a very primitive example. Number 001 or 100 was probably the first, starting in early '46. Over a dozen examples show a consistent correlation between serial numbers and features: higher numbers have later features.

By fall of '47, serial number 1000 was reached. The model continued until it was replaced by the tweed Deluxe, c. spring of '48. The wooden Professional, which is seen so rarely that it was probably custom-ordered, did not have its own control panel; the panel from the "Model 26" was used, sometimes with the model number scratched out. The Professionals possibly used the higher-numbered panels, as few with numbers under 1000 have turned up, and the amp was probably only made from mid '46 to early '47.

The tweeds can be dated by their style of tweed, in that models with unfinished tweed material have lower serial numbers than those with finish. Two-digit serial numbers stamped into the chassis of early Supers (by fall of '47) apparently were a new series and not a continuation of the Dual Professional (also with two-digit serial numbers). Numbers reached 400 by spring of '48 and 750 by fall of '49.

The tweed Princeton and Deluxe both reached the 1500 mark c. 1950. Champion "800s" seem to run from two-digit numbers through at least 700 during their short time from summer of '48 to spring of '49. Champion "600" numbers started over. Early Bassmans all seem to lack tube charts, but came with a four-digit serial number starting with 0 (probably 0001, 0010, or 0100 was first).

There you have it—a bunch of numbers and the most logical speculation musterable. It is quite possible that not all the numbers were used. This could have been to give the impression of higher production. If this is true, then there are a lot fewer of these around than the numbers indicate. And if they indeed ran in a strict sequential numbering scheme, then the production totals could be deduced. Someone should compile a gigantic list....

CHRONOLOGY OF THE EARLIEST FENDER AMPS

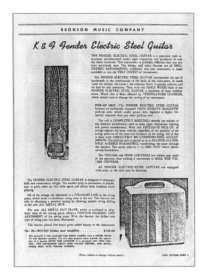

From the Radio-Tel foldout flyer, 1946.

From Radio-Tel catalog #151-A.

Although no promotional material has turned up on K&F, a promotional page for Bronson Music in faraway Detroit shows the transition from K&F to Fender Electric Instrument Co. A "K&F Fender Electric Steel Guitar" was offered; the photo clearly shows a late K&F model with rounded sides, a rectangular pickup/control assembly, and the extended chrome fingerboard. To go along with the instrument (total price for the pair: $150) was clearly a Fender Deluxe amp, a.k.a. the wooden "Model 26." (The number 26 probably referred to February 1946, when K&F became Fender.) The caption accompanying the picture described production models to a T. This must have been spring into summer of '46.

A photo of the original three Fender "woodies" in an early foldout flyer from Radio-Tel (Fender's distributors) showed what a second brochure, catalog #151-A, would label as the Princeton, Deluxe, and Professional models. (This brochure also included the new Professional Double Neck Guitar, dating it a bit later than the small foldout [c. late '46–early '47].) The Deluxe is by far the most common of the three today, followed by the Princeton and the extremely rare Professional.

According to Forrest White in his book *Fender: The Inside Story*, the Double Neck Steel Guitar was first made in early 1947 for Noel Boggs (with Bob Wills), and by early '47 there were two new tweed-covered amps: the Pro (as it was now called), with one 15" Jensen speaker, and the Dual Professional, with two 10" Jensens. (The small foldout brochure from Radio-Tel shows that the original trio of wooden amps was around before the Double Neck Steel, while "catalog #151-A" and an undated price list show that the Double Neck was around before the Dual Professional.)

One topic regarding the Dual Professional that has been misconstrued into being accepted as fact stems from a remark made about 12" speakers becoming more popular than 10s. The story now has it that Dual Professionals were made to use up or take advantage of excess 10" speakers bought at surplus prices. Not true at all! That Fender purchased a quantity of Jensen PM10-C speakers (with date codes of 220632, indicating the thirty-second week of 1946) for use in his Deluxe amps seems reasonable. As business was slow, they didn't get used as quickly as perhaps he had thought they would—a potential source of resentment towards them in later recollections. As for the 12s, Fender had never used them, and wouldn't start using them until the change from the wooden Deluxe to tweed more than a year later. Fact is, Jensen was a major manufacturer that had been in business since at least the early thirties and had thrived on defense work during the war, so whether Fender was buying its speakers direct from Jensen or from a distributor, it's doubtful anyone was clearing them out. (Almost all guitar amp manufacturers in the late forties—Gibson, National, Danelectro, Premier, et al.—used them.)

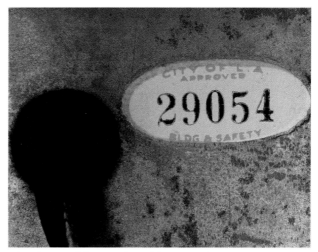

City of L.A. Building & Safety sticker. *From the inside of a wooden Princeton. Early Fender amps were given these stickers in compliance with municipal bureaucracy. Some say they can be used to date the amps.*

Let's give the designers some credit: The angled front on the Dual Professional was not done for looks, but to satisfy players' requests for better dispersion. Two 10" silver-framed Jensens would give a power-handling capability similar to that of the blue-framed Concert Series 15" at a total lower cost! Since the two-6L6 circuit would eat most single speakers of the day, this was a major concern for an amp that would be used professionally. The 10" speakers, as used for the wooden Deluxe, would have been the only practical speakers on hand, since the Princeton, Deluxe, and Professional used 8", 10", and 15" speakers, respectively. The "head" of the Dual Professional was essentially that of the Professional, so the single 15" was out. The use of more-expensive dual 12" speakers would not become necessary until the release of the higher-power Twin Amp more than five years later. The Dual Professional was and is a great design, logical in theory and practical in execution. Fender would become famous for this.

Two early Dual Professionals. *The "white" vertical weave on the left marked Fender's first use of tweed on an amp, followed shortly by the vertical tweed with contrasting threads on the right.*

White claims to have seen a Pro with a top control panel and the new "Fender/Fullerton, California" badge in early March 1947, so it's possible the tweed Pro was around at the same time as the first Dual Professionals. But it appears that the first tweed amp was the Dual Professional, since early examples were covered in a vertical-pattern monochromatic tweed, which looks almost white. (Has anyone seen a Pro in this material?) Later Dual Professionals and early tweed Pros were covered in unfinished vertical tweed, but a two-tone variety. The Dual Professional also introduced "finger-joint construction" cabinets and the use of a circuit board. The circuit board was used on all the tweed amps and the last half of the wooden Deluxe's run. This also reinforces dating the first tweed Pro to early '47, as seemingly all of the wooden Professionals lack the circuit board. These boards greatly improved reliability and the ease of manufacturing, so it would be hard to imagine that Fender would not have instituted their use for the top-of-the-line Pro.

Dating the tweed Pro to early '47 would explain the scarcity of the wooden Professionals as compared to the wooden Deluxes, which were still being sold into April of '48 before changing to tweed. (This extended run could have been to use up existing parts and materials.) The early tweed Pros were equipped with field-coil speakers, another point suggesting the early-'47 release, as Fender had discontinued the use of field-coil speakers on all its models as early as it could.

BIBLIOGRAPHY

The following books were used as reference material and are recommended reading.

For an in-depth look at the company, check out two books recently written by long-time Fender employees: *Fender: The Inside Story*, by Forrest White (Miller Freeman Books, 1994), and *Guitar Legends*, by George Fullerton (Centerstream, 1993). *American Guitars*, by Tom Wheeler (Harper & Row, 1982), you should have by now; if not, put it on your list. Also look for Richard Smith's book on Fender, coming soon.

For those interested in electronics, two "vintage" books published by Howard W. Sams & Company and still in print are: *Basic Electronics*, by Van Valkenburgh, Nooyer & Neville, Inc. (1955), and *How to Read Schematics*, by Donald Herrington (1962). These are good starters, particularly *Basic Electronics*, which was written for the U.S. Navy and explains tubes and tube amps.

Two books from the tube era that are no longer in print, but can be found on the used-book market, are *The Radiotron Designer's Handbook*, Fourth Edition (1952) and the *Audio Cyclopedia*, First (1959) or Second (1969) Edition. Expect to pay between $50 and $200 for these 1,000-plus-page cyclopedias.

The library of every amp aficionado should include at least one *RCA Receiving Tube Manual*, which includes info on all the tubes plus basic circuit theory and practical examples—rumored to have been influential on early Fender amp designs. Different eras contain different tubes, so look for more than one. (Editions from 1940 through the sixties were consulted for this book.) Expect to pay $5 to $20 for originals. The '59 edition has been reprinted and is available from Mojo Musical Supply, Napa, California. They have also compiled many of the pre-CBS schematics and layouts. These are reproduced at full size and placed in chronological order by model. Three volumes: Tweeds, Brown and White Tolex, and Blackface.

For a *really* in-depth look, find a college textbook on tubes from the original era. Heavy on math, but enlightening. For example, *Theory and Applications of Electron Tubes*, by Herbert J. Reich, Ph.D. (1939, 1944).

Back to Fender amps: The company published catalogs of their products on a regular basis, and these played an important part in the research for this book. They're also fun to collect! (Many of these have been reprinted by John Gima Reprints, Vintage Paper, and American Guitar Center [John Sprung's mail-order business].) Here's a fairly complete list: **1946:** small foldout; **c. 1947:** Radio & Television Equipment Co. catalog #151-A; **1948:** small full-line catalog; **1949:** large full-line catalog; **1950:** similar to 1949 plus Esquire; **1951:** N/A; **1952:** small foldout; **1953:** small foldout; **1954:** small foldout and large full-line catalog; **1955–65:** *Down Beat* insert and large full-line catalog, except no full-line in 1959; **1966:** small pamphlet and large full-line catalog; **1967:** N/A; **1968:** mini and large full-line catalogs (actually 1967–68); **1969, 1970, 1972, 1976, 1982:** large full-line catalogs; **1983:** amps brochure; **1987:** tube amps brochure; **1989, 1991, 1993, 1995:** amps brochures.

Also used were price lists from 1947 through 1995 (see "Availability Charts, with Prices").

INDEX

"II" series, 37, 107, 207
15, 153
75/30/140, 21, 132–134
85, 153
300 PS, 21, 130
400 PS Bass, 21, 122–124
availability charts, 238–243
B-300, 149
Bandmaster, 28, 89–93
Bandmaster Reverb, 93
Bantam Bass, 120
Bass control, 27, 80, 89, 222
bass preamp, 136
Bassman, 18, 27, 28, 29, 70–79, 182, 185
    solid-state, 144–145, 153
Bassman 10, 78
Bassman 20, 79
Bassman 50, 78, 79
Bassman 70, 79
Bassman 100, 78–79
Bassman 120, 150
Bassman 135, 79
Bassman Compact, 79, 150
Bassman Reissue, 79
bias, 220–221
blackface, 21, 32, 34, 198–199
    second-version, 37, 206
Blues Deluxe, 138
Blues DeVille, 138
Blues Junior, 138
Boost switch, 226
Bright switch, 75–76, 223
Bronco, 118–119, 138
    solid-state, 154
Bullet, 154
BXR series, 152–154
car battery, 151
CBS, 20–22, 35, 37, 226, 239, 241
Champ, 17, 19, 66–69
Champ II, 68
Champ 12, 39, 69
Champ 25 and 25SE, 69, 153–154
Champion 110, 153–154
Champion "800" and "600," 18, 27, 66–67, 180–181
Champs, the, 30
channel patching, 147
channel switching, 21, 132–133
chorus pedal, 175
circuit board
    first, 26, 59, 250
    lack of, 43, 48, 53, 67
compression, 53, 72, 219–220
compressor pedal, 175
Concert, 30, 106–109
control panel, 233
    front-facing, 29, 102
    separate from chassis, 70–72
Corona, California, 22
coverings, 69, 212–213. *See also* Tolex; tweed
    alligator skin, 36, 148
    carpet, 40, 153
    leatherette, 27
    lizard skin, 40, 141
    snakeskin, 69
Cox, Roger, 22
Custom Amp Shop, 22, 40, 139–141, 211
"Custom" Tweed Reverb, 142, 161
"Custom" Vibrasonic, 105, 142
"Custom" Vibrolux Reverb, 101, 142
date codes, 76, 237, 244–245
DeArmond accessories, 156
Deluxe, 19, 25, 26, 42–47, 53
Deluxe 85, 153, 248, 250
Deluxe 112, 47, 153–154
Deluxe 185, 47
Deluxe Reverb, 46–47
    solid-state, 145
Deluxe Reverb II, 47
Deluxe Reverb Reissue, 47
digital delay pedal, 175
Dimension IV, 146–147, 170
distortion, 21, 132–133
Distortion control, 128
distortion pedal, 171, 173, 175
Doppler effect, 167–168
Double Showman, 112
Dual Professional
    1940s, 25–26, 59–60, 179, 248–250
    Custom Amp Shop, 141
Dual Showman, 39, 112–114
    solid-state, 144–145
Dual Showman Reverb, 113–114
Dual Showman SR, 114
EccoFonic, 158
echo, 158, 163, 165, 166, 170, 172
Echo-Reverb, 165
Echoplex, 158, 165
effects, 36, 156–175, 243
Electronic Echo Chamber, 163
Electronic Echo Chamber Disc Delay, 165
equalization, 127–128, 223
Esquire, 18
Evans, Dick, 21
Fat switch, 105
Fender, Leo, 16–20, 24–25
Fender Blender, 173
Fender Electric Instrument Co., 16, 25, 35, 238
Fender Musical Instruments (CBS). *See* CBS
Fender Musical Instruments Corp. (FMIC), 22–23, 240, 242

Fender Sales, 17, 18–20
Fender's Radio Service, 16, 17
flanger pedal, 175
FMIC. *See* Fender Musical Instruments Corp.
Fullerton, George, 17, 102
Fuzz-Wah, 171
glides, 236
grille cloth, 28, 214–215
Grom, Steve, 22
Gulbransen Organ, 21
Haigler, Bob, 21, 22
Hall, F.C., 17, 18, 19
Hammond reverb, 159, 164
handles, 235
Harvard, 28, 97–98
   solid-state, 150
Harvard Reverb, 150
Harvard Reverb II, 151
Hayes, Charlie, 18, 19
Hoopeston, Illinois, 21
Hughes, Bill, 22
Hum Balance control, 226
Hyatt, Dale, 17
hybrid amps, 69, 153–154
input circuit, standard, 61
Jahns, Ed, 21, 47
K&F, 16, 24–25, 177
Kauffman, Doc, 16, 18, 25
keyboard amps, 152–154, 242
knobs, 234
KXR series, 154
Lake Oswego, Oregon, 22
Leslie speaker, 167–168
London 185, 153
London Reverb, 151
Lopez, Lupe, 17
Lover, Seth, 21, 146
Lugar, Louis, 17
M-80 series, 40, 153
maintenance, 227–229
Marshall, 73
Massie, Ray, 17
Master Volume, 78–79, 85, 132–133, 226, 229
Middle control, 73, 82, 223
mini amp, 39
Mix control, 109, 140
"Model 26," 25, 42, 53, 248
model names, confusing, 34, 57, 62
model numbers, 237, 246
modifications, 229–230
Montreux, 151
Multi-Echo, 172
Musicmaster Bass, 121
nameplates, 232
narrow-panel, 28, 184, 186–187
negative feedback, 222–223

New Vintage series, 142
   "Custom" Tweed Reverb, 161
   "Custom" Vibrasonic, 105
   "Custom" Vibrolux Reverb, 101
Orchestration +, 169
Ortega, Maybelle, 17
Performer series, 153–154
Perkins, Cal, 22
phase inverter, 121, 221–222
Phaser, 174
piggyback, 30–31, 110–111, 185, 190
pot codes, 245
Power Chorus, 153
Precision Bass, 18, 70, 71
Presence control, 223
price charts, 238–243
price increase of 1970, 37, 239, 241, 243
Princeton, 19, 25, 48–52, 248
Princeton 112, 52, 153–154
Princeton Chorus, 52, 153–154
Princeton Reverb, 50, 51
Princeton Reverb II, 52
Pro, 19, 26, 27, 53–58, 248, 250
Pro 185, 58, 153
Pro Junior, 137–138
Pro Reverb, 57–58
   solid-state, 145
Professional, 25, 26, 53–54, 248, 250
"Professional" series (1960s), 30, 104
Professional Tube series (1990s), 108, 211
Prosonic, 141
pull-knob switch, 223, 226
push-pull amplification, 221–222
Quad Reverb, 126
rack-mount, 135, 149
R.A.D./H.O.T./J.A.M., 40, 153–154
Radio & Television Equipment Co. (Radio-Tel), 17, 18, 19, 248
Randall, Don, 17, 18–20
"Red-Knob" series, 39, 114, 208
reissue guitars, 21, 39
Reissue series, 39, 209
   '59 Bassman, 79
   '63 Fender Reverb, 161
   '63 Vibroverb, 101, 116–117
   '65 Deluxe Reverb, 47
   '65 Twin Reverb, 87
reverb, 32, 116, 159
Reverb jacks, 170, 172
Reverb Unit, 159–161, 200. *See also* Solid-State Reverberation Unit
RGP-1, 135–136
Rickenbacker, 17, 19
Rissi, Bob, 21
Rivera, Paul, 21, 37, 47
rotating speaker, 167–168
RPW-1, 135–136
Rumble Bass, 140

INDEX

sag, 72
Sanchez, Lydia, 17
Schultz, William, 7
Scottsdale, Arizona, 23
serial numbers, 245–247
series-parallel, 65, 125
Showman, 31, 110–114
   solid-state, 114, 151
Sidekick series, 38, 151, 152–153
Sidekick Bass series, 150, 152–153
silverface, 21, 36–37, 203–204, 245
   changing circuit to blackface, 63, 229–230
solid-state amps, 144–154, 241–242
   first-series, 20–21, 35–36, 144–145, 201
   second-series, 21, 37–38, 150–151
   third-series, 22, 39, 152–154
Solid-State Reverberation Unit, 164
Soundette, 166
speaker cabinet
   folded horn, 122
   ported, 70, 110
   powered, 146
speaker jacks, alternating, 123
speakers
   alignment, 28, 61, 63, 81–83
   angled, 130
   date codes, 245
   dual, 59, 60, 249
   early, 249
   field-coil, 25, 250
   Styrofoam, 120
Special Effects Center, 170
Spranger, Paul, 21
Squier 15, 152
Stage 112SE, 153–154
Stage 185, 153
Stage Lead, 151
steel guitar, 16, 17, 18
stereo chorus pedal, 175
stomp boxes, 175
Stratocaster, 18, 19
Student amp, 67
Studio Bass, 131
Studio Lead, 151
Style switch, 144
Sunn, 22
Super, 26, 27–28, 59–65
Super 60, 64–65
Super 112, 64–65
Super 210, 64–65
Super Bassman, 78
Super Champ, 68
Super Champ Deluxe, 68
Super Reverb, 62–64
   solid-state, 145
Super Showman, 146–147, 170
Super Six Reverb, 125
Super Twin, 127–129

Super Twin Reverb, 128–129
swamper resistor, 73
Sweet switch, 105
T-nuts, 75, 111
Tavares, Freddie, 19, 21
Tel-Ray Electronics, 170
Telecaster, 18
"The Twin," 39, 86–87
tilt-back legs, 75, 110
Tolex, 29, 188–197
Tone control, 80, 222
tone ring, 110
Tonemaster, 139
TR 105, 162
Treble control, 27, 80, 89, 222
tremolo, 28, 94, 103, 224–225
Tremolux, 19, 28, 94–96
Tualatin, Oregon, 22
tube amps, 21, 23, 42–142, 238–240
tube charts, 66, 237, 244, 246
tubes, 218–220
   esoteric, 129, 219
TV-front, 26, 28, 180–181
tweed, 25, 26, 27, 60, 250
Tweed series, 40, 137–138
Twin, 18, 22, 27–28, 29, 80–88
Twin, The. See "The Twin"
"Twin-Amp," 87–88
Twin Reverb, 84–86
   solid-state, 144
Twin Reverb II, 86
Twin Reverb Reissue, 87
Ultra Chorus, 153–154
V-front, 26
Variable Echo-Reverb, 165
Vibrasonic, 29, 102–105, 188
vibrato, 28, 103
Vibratone, 167–168
Vibro Champ, 68, 118–119
Vibro King, 139
Vibrolux, 28, 99–101
Vibrolux Reverb, 100–101
   solid-state, 145
Vibrosonic Reverb, 104–105
Vibroverb, 32, 101, 115–117, 193
Vibroverb Reissue, 101, 116–117
Volume Pedal and Volume Tone Pedal, 156–157
wah-wah, 171
Wentling, Mark, 22
White, Forrest, 17, 19, 29, 248, 250
"White" amps, 29, 185
wide-panel, 27, 28, 183
Wills, Bob, 25
wireless microphone, 162
"woodies," 17, 25, 26, 178, 248, 250
Yale Reverb, 151
Zodiac amps, 21, 36, 148, 202

ABOUT THE AUTHORS

(Photo by Bob Basone)

From tweed Champs and Deluxes to white Tolex Showmans and Bassmans, from blackface Vibrolux Reverbs and two colors of Vibroverbs to a silverface Super Six Reverb and a brand-spankin'-new Vibro King, JOHN TEAGLE has spent a great portion of his adult life at the controls of a Fender amp. Twenty years of playing in bands, working in music stores, engineering and producing recording sessions, running sound for bands, teaching, repairing, and writing about guitars, and studying electronics went into preparing him for writing this book. A fanatical collector of "paper," particularly old musical instrument catalogs, he lives in New York City with his wife and infant daughter. Spare time is spent playing in two bands: the Vice Royals (recently chosen Best NYC Instrumental Rock Band by NY Press) and the surf guitar instrumental combo Purple kniF.

JOHN SPRUNG's interest in vintage Fender equipment was sparked by an early experience with a 1952 Precision Bass, which set him off in search of a '51. The fire was fanned some years later by an encounter with a wooden Deluxe "Model 26" amp. With the help of Fender expert Cesar Diaz, Sprung began collecting pre-1950 Fender amps with a vengeance. Then the photo bug bit. The two hobbies went well together, and it wasn't long before Sprung's photos began appearing in magazines and books.

(Photo by Jenny Sprung)

The association between Teagle and Sprung began in the mid eighties. Their common interest in Fender amps and old catalogs kept them in contact for the next 10 years. Both now serve as senior contributing editors for *20th Century Guitar* magazine. Sprung's knowledge of the early Fender amps, combined with Teagle's vast experience with the tweed and Tolex-era amps, made them the logical partnership for the production of this book.